T0220401

Bandwidth Extension of Speech Using Perceptual Criteria

Synthesis Lectures on Algorithms and Software in Engineering

Editor
Andreas Spanias, *SenSIP Center, School of ECEE, Arizona State University*

Bandwidth Extension of Speech Using Perceptual Criteria
Visar Berisha, Steven Sandoval, and Julie Liss
2013

Control Grid Motion Estimation for Efficient Application of Optical Flow
Christine M. Zwart and David H. Frakes
2013

Sparse Representations for Radar with MATLAB® Examples
Peter Knee
2012

Analysis of the MPEG-1 Layer III (MP3) Algorithm Using MATLAB
Jayaraman J. Thiagarajan and Andreas Spanias
2011

Theory and Applications of Gaussian Quadrature Methods
Narayan Kovvali
2011

Algorithms and Software for Predictive and Perceptual Modeling of Speech
Venkatraman Atti
2011

Adaptive High-Resolution Sensor Waveform Design for Tracking
Ioannis Kyriakides, Darryl Morrell, and Antonia Papandreou-Suppappola
2010

MATLAB® Software for the Code Excited Linear Prediction Algorithm: The Federal
Standard-1016
Karthikeyan N. Ramamurthy and Andreas S. Spanias
2010

OFDM Systems for Wireless Communications
Adarsh B. Narasimhamurthy, Mahesh K. Banavar, and Cihan Tepedelenliouglu
2010

Advances in Modern Blind Signal Separation Algorithms: Theory and Applications
Kostas Kokkinakis and Philipos C. Loizou
2010

Advances in Waveform-Agile Sensing for Tracking
Sandeep Prasad Sira, Antonia Papandreou-Suppappola, and Darryl Morrell
2008

Despeckle Filtering Algorithms and Software for Ultrasound Imaging
Christos P. Loizou and Constantinos S. Pattichis
2008

Bandwidth Extension of Speech Using Perceptual Criteria
Visar Berisha, Steven Sandoval, and Julie Liss

ISBN: 978-3-031-00393-6   paperback
ISBN: 978-3-031-01521-2   ebook

DOI 10.1007/978-3-031-01521-2

A Publication in the Springer series
SYNTHESIS LECTURES ON ALGORITHMS AND SOFTWARE IN ENGINEERING

Lecture #13
Series Editor: Andreas Spanias, *SenSIP Center, School of ECEE, Arizona State University*
Series ISSN
Synthesis Lectures on Algorithms and Software in Engineering
Print 1938-1727   Electronic 1938-1735

Bandwidth Extension of Speech Using Perceptual Criteria

Visar Berisha, Steven Sandoval, and Julie Liss
Arizona State University

SYNTHESIS LECTURES ON ALGORITHMS AND SOFTWARE IN ENGINEERING #13

ABSTRACT

Bandwidth extension of speech is used in the International Telecommunication Union G.729.1 standard in which the narrowband bitstream is combined with quantized high-band parameters. Although this system produces high-quality wideband speech, the additional bits used to represent the high band can be further reduced. In addition to the algorithm used in the G.729.1 standard, bandwidth extension methods based on spectrum prediction have also been proposed. Although these algorithms do not require additional bits, they perform poorly when the correlation between the low and the high band is weak. In this book, two wideband speech coding algorithms that rely on bandwidth extension are developed. The algorithms operate as wrappers around existing narrowband compression schemes. More specifically, in these algorithms, the low band is encoded using an existing toll-quality narrowband system, whereas the high band is generated using the proposed extension techniques. The first method relies only on transmitted high-band information to generate the wideband speech. The second algorithm uses a constrained minimum mean square error estimator that combines transmitted high-band envelope information with a predictive scheme driven by narrowband features. Both algorithms make use of novel perceptual models based on loudness that determine optimum quantization strategies for wideband recovery and synthesis. Objective and subjective evaluations reveal that the proposed system performs at a lower average bit rate while improving speech quality when compared to other similar algorithms.

KEYWORDS

Speech compression, bandwidth extension, loudness, psychoacoustics

Contents

Acknowledgments

The work in this book was supported in part by the SenSIP Center, Arizona State University, the NIH NIDCD R01 and R21 grants, and the NSF PhD Fellowship program.

Visar Berisha, Steven Sandoval, and Julie Liss
September 2013

Figure Credits

Figure 3.1   from Fletcher, H. Perceptual segmentation and component selection for sinusoidal representations of audio. *Reviews of Modern Physics*, vol. 12, pages 47–65. Copyright © 1940 by The American Physical Society. 10.1103/RevModPhys.12.47

CHAPTER 1

Introduction

Speech coding is the field concerned with reducing the bit rate representation of voice signals [1]. Voice coders or *vocoders* are algorithms used for data rate reduction. At the early stages of speech coding research, algorithms were primarily targeted for military communications and encryption applications that focused primarily on low bit rate, narrowband representations of speech [1]. As digital communications started gaining ground in telephony applications [2], the bandwidth conservation and enhanced privacy aspects of this technology created a lot of interest in industry circles. Work in the area started with analog channel vocoders and later on Differential Pulse Code Modulation (DPCM) systems used in digital telephone switches [1]. The emergence of DSP chips [3] and the development of linear predictive coding methods for ADPCM [2] and for more elaborate analysis-synthesis systems were important milestones in this area. The emergence of cellular telephony in the 1980s [4] propelled research in speech coding to new levels; as a result, a series of sophisticated narrowband vocoder algorithms began finding their way into products and telephony standards [5, 6]. The international telecommunications union (ITU) and the telecommunication industry association (TIA) were instrumental in rapidly establishing compatibility standards for wireless and wireline telephony. The Department of Defense also proposed linear prediction standards for military applications [95]. The computer and multimedia industry also adopted several vocoder algorithms for streaming and voice over internet protocol (VoIP) applications.

The emergence of third-generation cellular networks in the 2000s has created new enhanced mobility applications with promise to accommodate wideband and multirate speech coders. We have recently witnessed a renewed interest in speech processing to develop a new generation of wideband algorithms or to retrofit existing narrowband standards with bandwidth extension capabilities. Examples of such algorithms include the ITU G.729.1 compression method. In this standard, the low band (300–3.4 kHz) is encoded using an existing narrowband coder (ITU G.729), whereas the high band (3.4–7 kHz) is coarsely parameterized and extended using fewer bits. In this book, we present an in-depth study of methods for high-band bandwidth extension and discuss perceptual criteria for optimization techniques.

1.1 MOTIVATION

The public-switched telephony network (PSTN) and most of today's cellular networks use narrowband (0.3–3.4 kHz) speech coders. This, in turn, places limits on the naturalness and intelligibility of speech [1] and is most problematic for sounds whose energy is spread over the entire audible spectrum. For example, unvoiced sounds such as "s" and "f" are often difficult to discriminate with a narrowband representation. In Fig. 1.1, we provide spectral plots of different phonemes. For the fricatives ("s", "sh", "z"), the energy is spread throughout the spectrum, however most of the energy of the vowels ("ae", "aa", "ay") lies within the low-frequency range [7]. Split-band compression algorithms recover the narrowband spectrum (0.3–3.4 kHz) and the high-band spectrum (3.4–7 kHz) separately [8, 9]. The main goal of these algorithms is to produce wideband (0.3–7 kHz) speech through the use of bandwidth extension. In other words, an existing narrowband encoder compresses the low-band signal, whereas the high band is extended using fewer bits. A number of these techniques make use of the correlation between the low band and the high band to predict the wideband speech from narrowband features [10–14]. Some of these algorithms attempt to cleverly embed the high-band parameters in the low-frequency band [15, 16]. Others generate coarse representations of the high band at the encoder and transmit them as side information to the decoder [8–10, 17–20]. Bandwidth extension algorithms either predict the

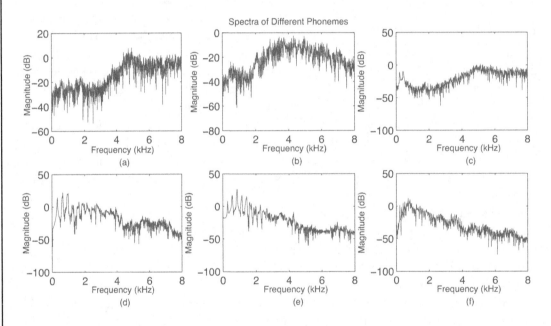

Figure 1.1: The spectra of six different phonemes. More specifically, the spectra for "s" (a), "sh" (b), "z" (c), "ae" (d), "aa" (e), "ay" (f). Most of the energy of the fricatives ((a), (b), and (c)) falls in the high band and therefore a wideband representation is required for these phonemes.

high band or attempt to coarsely parameterize it. Recent studies, however, show that the mutual information between the narrowband and the high-frequency bands is insufficient for wideband synthesis solely based on prediction [21–23]. In fact, Nilsson *et al.* [23] show that the available narrowband information reduces uncertainty in the high band, on average, by only ≈ 10%. As a result, some side information must be transmitted to the decoder in order to accurately characterize the wideband speech. Existing algorithms typically transmit side information for every frame, regardless of how much that particular frame benefits from a wideband representation. This is because today's bandwidth extension algorithms do not rely heavily on psychoacoustic criteria.

Most existing wideband recovery techniques are based on the source/filter model [10, 12, 13, 24]. These techniques typically include implicit psychoacoustic principles, such as perceptual weighting filters and dynamic bit allocation schemes in which lower-frequency components are allotted a larger number of bits. Although some of these methods were shown to improve the quality of the coded audio, studies show that additional coding gain is possible through the integration of explicit psychoacoustic models [25–28]. Existing psychoacoustic models are particularly useful in high-fidelity audio coding applications, however their potential has not been fully utilized in traditional speech compression algorithms or wideband recovery schemes.

1.2   EXISTING BANDWIDTH EXTENSION ALGORITHMS

Most bandwidth extension algorithms fall into one of two categories: bandwidth extension based on explicit high-band generation and bandwidth extension based on the source/filter model. Early methods for artificial bandwidth extension involved band replication followed by spectral shaping [29–31]. For example, consider a narrowband signal resampled at 16 kHz, $s_{nb}(k)$. To generate an artificial wideband representation, the signal is first upsampled as shown in (1.1). The upsampler folds the low-band spectrum (0–4 kHz) onto the high band (4–8 kHz) and therefore fills out the entire spectrum.

$$\hat{s}_{1,wb}(k) = \begin{cases} s_{nb}(k/2) & \text{if } mod(\frac{k}{2}) = 0, \\ 0 & \text{else.} \end{cases} \tag{1.1}$$

Following the spectral folding, the high band is transformed by a shaping filter, $s(k)$, as shown in (1.2).

$$\hat{s}_{wb}(k) = \hat{s}_{1,wb}(k) * s(k) * g(k). \tag{1.2}$$

Different shaping filters are typically used for different frame types. For example, the shaping associated with a voiced frame may introduce a pronounced spectral tilt, whereas the shaping of an unvoiced frame tends to maintain a flat spectrum. Following the high-band shaping, a gain control mechanism, $g(k)$, controls the gain of the low band and the high band such that their relative levels are suitable.

Examples of techniques based on similar principles include [29], [30], and [31]. Although these simple techniques can potentially improve the quality of the speech, audible artifacts are often induced. The artifacts are most often introduced for frames when the high band greatly differs from the low band. In such scenarios, spectral folding techniques create artificial wideband representations with significantly different spectral structures when compared to the original wideband audio.

As a result of the limitations discussed above, more sophisticated techniques based on the source/filter speech production model have been developed [10–13, 32]. The auto-regressive (AR) model for speech synthesis is given by:

$$\hat{s}_{nb}(k) = \hat{u}_{nb}(k) * \hat{h}_{nb}(k), \tag{1.3}$$

where $\hat{h}_{nb}(k)$ is the impulse response of the filter given by $\hat{H}_{nb}(z) = \frac{\sigma^2}{\hat{A}_{nb}(z)}$. $\hat{A}_{nb}(z)$ is a quantized version of the Nth order LP filter given by:

$$A_{nb}(z) = \sum_{i=0}^{N} a_{i,nb} z^{-i}, \tag{1.4}$$

and $\hat{u}_{nb}(k)$ is a quantized version of

$$u_{nb}(k) = s_{nb}(k) - \sum_{i=1}^{N} a_{i,nb} s_{nb}(k-i). \tag{1.5}$$

A general procedure for performing wideband recovery based on the speech production model is given in Fig. 1.2 [32]. In general, a two-step process is taken to recover the missing band. The first step involves the estimation of the wideband source-filter parameters, a_{wb}, given certain features extracted from the narrowband speech signal, $s_{nb}(k)$. The second step involves extending the narrowband excitation, $u_{nb}(k)$. The estimated parameters are then used to synthesize the wideband speech estimate. The resulting speech is high-pass filtered and added to a 16 kHz resampled version of the original narrowband speech, denoted by $s'_{nb}(k)$, as shown in Equation (1.6).

$$\hat{s}_{wb}(k) = s'_{nb}(k) + g_{HPF}(k) * [h_{wb}(k) * u_{wb}(k)] . \tag{1.6}$$

This approach has been used in a number of different algorithms [12, 32–38]. In [33] and [34], the authors make use of dual, coupled codebooks for parameter estimation. In [12], [35], and [36], the authors use statistical recovery functions that are obtained from pre-trained Gaussian mixture models (GMM) or hidden Markov models (HMM). Yet another set of techniques use linear wideband recovery functions [37, 38]. The underlying assumption for most of these approaches is that there is sufficient correlation between the narrowband features and the wideband envelope to be predicted. While this is true for some frames, it has been shown that the assumption does not hold, in general [21–23]. In fact, recent studies show that the mutual information between the narrowband and the high-frequency bands is insufficient for wideband synthesis solely based on prediction [21–23].

In Table 1.1, we show a measure of predictability for the high-band for two different scenarios developed by Nilsson *et al.* [23]. The predictability metric here is a ratio of the mutual information between a set of low-band and high-band features and the uncertainty (entropy) of the high-band features. In Table 1.1 (a) we show this normalized mutual information metric between the narrowband cepstrum and the high-band to low-band energy ratio. In Table 1.1 (b)

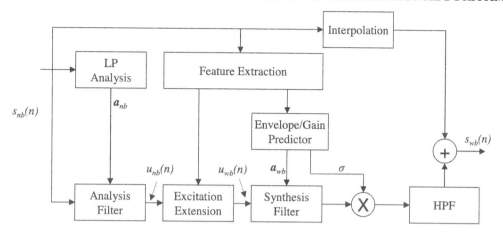

Figure 1.2: High-level diagram of traditional bandwidth extension techniques based on the source/filter model.[32]

we show the same metric between the narrowband cepstrum and the high-band cepstrum. As the tables show, the available narrowband information reduces uncertainty in the high-band energy by only ≈ 13% and in the high-band cepstrum by only ≈ 9%. These results imply that algorithms based on predicting the high-band often generate erroneous estimates [8]. It is therefore evident that for improved robustness the high-band spectrum must be quantized and transmitted as side information.

Table 1.1: (a) A ratio between the mutual information of the narrowband cepstral coefficients ($\mathbf{f}$) and the high-band energy ratio (y), $I(\mathbf{f})$, and the entropy of the high-band energy ratio, $H(y)$, for different sounds. (b) A ratio between the mutual information of the narrowband cepstral coefficients ($\mathbf{f}$) and the high-band cepstral coefficients ($\mathbf{y}$), $I(\mathbf{f};\mathbf{y})$, and the entropy of the high-band cepstral coefficients, $H(\mathbf{y})$, for different sounds.

sound	$I(\mathbf{f};\mathbf{y})$	$H(y)$	$\frac{I(\mathbf{f};y)}{H(y)}$
V	0.57	4.56	12.5%
G	0.59	4.78	12.34%
C	0.87	5.29	16.45%
S	0.56	4.97	11.27%
F	0.45	4.60	9.78%
N	0.55	4.63	11.88%

(a)

sound	$I(\mathbf{f};\mathbf{y})$	$H(\mathbf{y})$	$\frac{I(\mathbf{f};\mathbf{y})}{H(\mathbf{y})}$
V	1.45	14.77	9.82 %
G	1.53	13.80	11.09 %
C	1.60	14.83	10.79 %
S	1.23	15.55	7.91 %
F	1.09	15.02	7.26 %
N	1.20	15.82	7.59 %

(b)

In Fig. 1.3, we show examples of two frames that illustrate this point. The figure shows two frames of wideband speech along with the true and predicted envelopes. The estimated envelope was predicted using a technique based on coupled, pre-trained codebooks; a technique representative of most modern envelope extension algorithms [39]. Fig. 1.3 (top) shows a frame for which the predicted envelope matches the actual envelope quite well. In Fig. 1.3 (bottom), the estimated envelope greatly deviates from the actual and, in fact, erroneously introduces two high-band formants. In addition, it misses the two formants located between 4 kHz and 6 kHz. Consequently, a recent trend in bandwidth extension has been to transmit additional high-band information rather than using prediction models or codebooks to generate the missing bands [19].

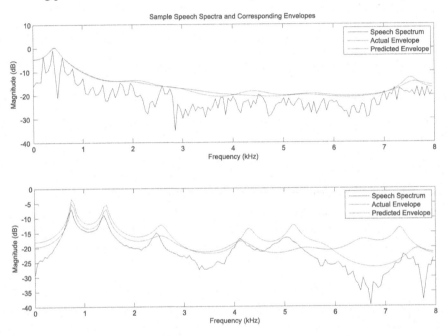

Figure 1.3: Wideband speech spectra (in dB) and their actual and predicted envelopes for two frames. The top figure shows a frame for which the predicted envelope matches the actual envelope. In the bottom figure, the estimated envelope greatly deviates from the actual.

A few split-band coders based on coarse high-band representations have been recently proposed [10, 18, 19, 40]. Although these techniques improve speech quality when compared to techniques solely based on prediction, psychoacoustic models are not fully exploited for high-band synthesis. In fact, the bit rates associated with the high-band representation are often unnecessarily high because they allocate the same number of bits for high band generation to each frame. It is apparent from Fig. 1.1 that a wideband representation is more beneficial for certain frame types (e.g., unvoiced fricatives). For a more quantitative analysis of the benefit of a wideband representation, please refer to Fig. 1.4. In this figure, we plot the estimated bandwidth of different phonemes over 100 s of speech obtained from the TIMIT database [41]. We define the bandwidth as the last frequency component to cross the mean of the fast Fourier transform

(FFT) magnitude. As the figure shows, several phonemes are adequately represented with an 8 kHZ sampling rate, the exception being some of the fricatives and the affricates. As such, algorithms that perform bandwidth extension by encoding the high band of *every* frame often operate at unnecessarily high bit rates.

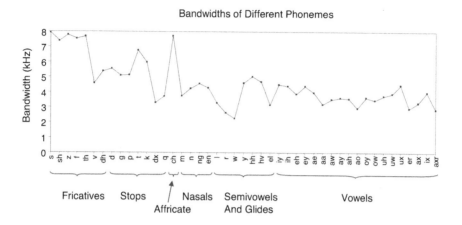

Figure 1.4: The estimated bandwidths of different phonemes.

Furthermore, consider the sonority scale shown in Table 1.2 [42]. This is a ranking of the amount of "sound" that a particular phoneme produces, with the obstruents producing the least and the sonorants producing the most. If we consider the amount of sound produced by a phoneme as a measure of perceptual relevance, the scale states that vowels are typically more important than plosives or fricatives. Looking at Table 1.2, vowels only require an 8 kHz sampling rate for capturing most of the energy, whereas the perceptually less relevant fricatives and plosives require 16 kHz sampling rates. This further motivates the need for a variable bit rate encoder that only performs bandwidth extension when there is a distinct perceptual gain.

Table 1.2: The sonority scale: a ranking of the amount of "sound" that phonemes produce, with 1 being the least and 6 being the most

Sonority Level	Phoneme Type	Obstruent/Sonorant
1	plosives	Obstruent
2	fricatives	Obstruent
3	nasals	Sonorant
4	liquids	Sonorant
5	high vowels	Sonorant
6	non-high vowels	Sonorant

Since the higher-frequency bands are perceptually less relevant, a coarse representation is often sufficient for a perceptually transparent representation [19, 40]. This idea is used in high-fidelity audio coding based on spectral band replication [40] and in the standardized G.729.1 speech coder [19]. Both of these methods employ an existing codec for the lower-frequency band while the high band is coarsely parameterized using fewer parameters. Although these techniques greatly improve speech quality when compared to techniques solely based on prediction, no explicit psychoacoustic models are employed for high-band synthesis. Hence, the bit rates associated with the high band representation are often unnecessarily high.

1.3 EXISTING PERCEPTUAL MODELS

Most existing wideband coding algorithms attempt to integrate indirect perceptual criteria in compression algorithms to increase coding gain. Examples of such methods include perceptual weighting filters [43], perceptual LP techniques [44], and weighted LP techniques [45]. The perceptual weighting filter attempts to shape the quantization noise such that it falls in areas of high signal energy, however it is unsuitable for signals with a large spectral tilt (i.e., wideband speech). The perceptual LP technique filters the input speech signal with a filterbank that mimics the ear's critical band structure. The weighted LP technique manipulates the axis of the input signal such that the lower, perceptually more relevant, frequencies are given a larger weight. Although these methods improve the quality of the coded audio, additional gains are possible through the integration of an explicit psychoacoustic model.

Over the years, researchers have studied numerous explicit mathematical representations of the human auditory system for the purpose of including them in audio compression algorithms. The most popular of these representations include the global masking threshold [46], the auditory excitation pattern (AEP) [47], and the perceptual loudness [48].

A masking threshold refers to a threshold below which a certain tone/noise signal is rendered inaudible due to the presence of another tone/noise masker. The global masking threshold (GMT) is obtained by combining individual masking thresholds; it represents a spectral threshold that determines whether a frequency component is audible [46]. The GMT provides insight into the amount of noise that can be introduced into a frame without creating perceptual artifacts. Psychoacoustic models based on the global masking threshold have been used to shape the quantization noise in standardized audio compression algorithms, e.g., the ISO/IEC MPEG-1 layer 3 [46], the DTS [49], and the Dolby AC-3 [50]. In Fig. 1.5 we show a frame of audio along with its GMT. The masking threshold was calculated using the psychoacoustic model 1 described in the MPEG-1 algorithm.

Auditory excitation patterns (AEP) describe the stimulation of the neural receptors caused by an audio signal. Each neural receptor is tuned to a specific frequency, therefore the AEP represents the output of each aural "filter" as a function of the center frequency of that filter; therefore, two signals with similar excitation patterns tend to be perceptually similar. An excitation pattern-matching technique called excitation similarity weighting (ESW) was proposed by Painter and

Spanias for scalable audio coding [51]. ESW was initially proposed in the context of sinusoidal modeling of audio. ESW ranks and selects the perceptually relevant sinusoids for scalable coding. The technique was then adapted for use in a perceptually motivated linear prediction algorithm [52].

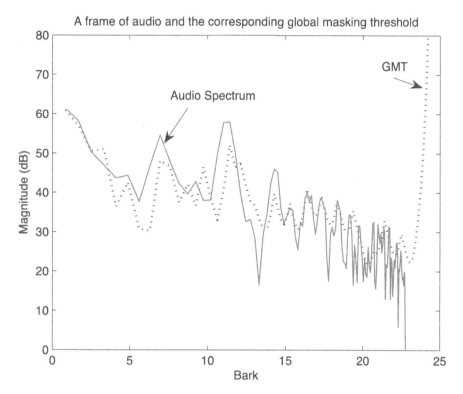

Figure 1.5: A frame of audio and the corresponding global masking threshold as determined by psychoacoustic model 1 in the MPEG-1 specification.

A concept closely related to excitation patterns is perceptual loudness. Loudness is defined as the perceived intensity (in sones) of an aural stimulation. It is obtained through a nonlinear transformation and integration of the excitation pattern [48]. Although it has found limited use in coding applications, a model for sinusoidal coding based on loudness was recently proposed [53]. In addition, a perceptual segmentation algorithm based on partial loudness was proposed in [51].

Although the models described above have proven very useful in high-fidelity audio compression schemes, they share a common limitation in the context of bandwidth extension. There exists no natural method for the explicit inclusion of these principles in wideband recovery schemes.

CHAPTER 2

Principles of Bandwidth Extension

In this chapter we discuss the basics of bandwidth extension and provide a literature survey of the area. Most bandwidth extension algorithms fall in one of two categories, bandwidth extension based on explicit high-band generation and bandwidth extension based on the source/filter model. Explicit high-band generation uses speech spectral translation or speech spectral folding followed by frequency scaling of the generated band. Extension algorithms based on the source/filter model extend the high-band envelope and the high-band excitation separately. This section describes the most popular techniques used in both categories, with a focus on techniques based on the source/filter model. The section begins with a description of a number of different techniques that predict the high band explicitly, followed by a discussion of bandwidth extension methods based on the source/filter model.

2.1 EXPLICIT HIGH-BAND GENERATION

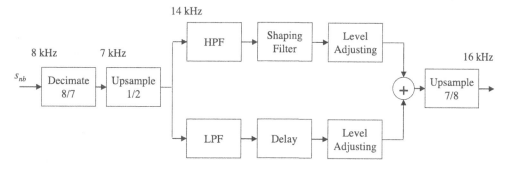

Figure 2.1: A bandwidth extension algorithm proposed by Yasukawa [29].

Early methods for artificial bandwidth extension involved spectral folding of the speech signal followed by a shaping filter. Yasukawa proposed a technique based on the block diagram shown in Fig. 2.1 [29]. As is seen in the block diagram, the original 8 kHz sampled narrowband speech is first decimated, then upsampled to 14 kHz, thereby creating a copy of the narrowband frequencies (0–3.5 kHz) in the high band (3.5–7 kHz). The resulting signal is then separated into high- and low-band frequencies by means of a low-pass and high-pass filter. The high-pass portion is transformed by the shaping filter and the levels of both the low-pass and high-pass portions are adjusted such that their relative levels are suitable. Once both the low-pass and high-pass versions of the speech signal are processed, the output is formed by summing the two bands.

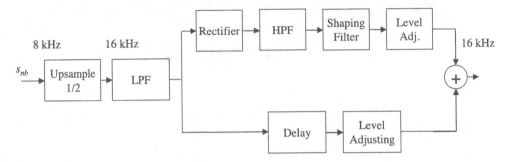

Figure 2.2: An update of the original bandwidth extension method proposed by Yasukawa. The added rectifier artificially generates high-band harmonics [30].

Further work was done by Yasukawa [30] and the original method was modified by including a rectifier before the shaping filter, as is seen in Fig. 2.2. The purpose of the rectifier is to artificially generate harmonics of the fundamental frequency. The performance of both of these techniques greatly depends on the choice of the shaping filter. The high-frequency envelopes vary from frame-to-frame, therefore the shaping filter must be designed such that it fits the characteristics of the frame being modeled.

Another method based on spectral folding is presented in [54]. The input signal, $s_{nb}(t)$, is first upsampled to a sampling frequency of 16 kHz and the aliased frequency components are generated in the high band from 4–8 kHz. Next, some time domain features are extracted from the original narrowband signal. These features are used as an input to a frame classifier that classifies each frame into three different categories: voiced sounds, sibilants, and stop consonants. The input features for this classifier are the following.

- The gradient index—a measure of the sum of the magnitudes of the gradient of the speech signal at changes of direction.

- The gradient count—the number of times the gradient index is greater than a predefined threshold.

- The energy ratio—the ratio between the energy of the current frame and that of the last frame.

- The narrowband slope—the slope of the narrowband speech frame between 0.3 and 3.0 kHz.

The computed features are compared against some pre-defined thresholds and the frame is classified into the appropriate group. The purpose for the classification is to construct spline functions used to shape the high band appropriately. Based on this classification and based upon the narrowband signal, a high-band shaping function is generated. This technique is representative of bandwidth extension methods that operate directly on the speech, rather than on the source/filter parameters [29–31].

Although these simple techniques can potentially improve the quality of the speech, oftentimes audible artifacts are induced. In the ensuing section we discuss methods of wideband recovery based on the source/filter model of speech. In these algorithms, the envelope and excitation are extended in frequency separately. A predictor is trained to generate the wideband envelope of a speech segment and the wideband excitation is artificially generated, as will be discussed later.

2.2 HIGH-BAND GENERATION BASED ON THE SOURCE/FILTER MODEL

Most of the successful bandwidth extension algorithms are based on the source/filter speech production model. In general, a two-step process is taken in order to recover the missing high-frequency band. The first step involves the estimation of the wideband source/filter parameters given certain features of the narrowband speech signal and some a priori knowledge about the properties and structure of wideband spectra. The estimated parameters are then used to synthesize the wideband speech estimate. This approach has been shown to be successful in a number of different algorithms [10–13, 32].

The general structure of source/filter bandwidth extension algorithms for speech is shown in Fig. 2.3. The AR coefficients of the wideband speech signal, $\mathbf{a}_{wb}$ and the high-band gain σ, are estimated from features extracted from the narrowband speech signal, $s_{nb}(t)$. The narrowband excitation, $u_{nb}(t)$, is extended using any number of techniques to be discussed in the ensuing section. The wideband excitation, $u_{wb}(t)$, then drives the wideband synthesis filter. The output of the synthesis filter is filtered and added to an interpolated version of the narrowband speech signal to form the wideband speech.

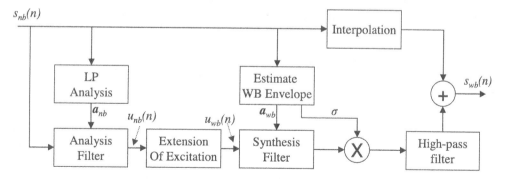

Figure 2.3: A block diagram of a typical bandwidth extension algorithm.[32]

As the figure shows, the algorithms treat the extension of the excitation and that of the spectral envelope separately since the two constituents can be assumed to be approximately independent [32]. In addition, by separately extending the source and the system, we can apply algorithms of differing complexity to the extension of each since listening tests have verified that for high-frequency bandwidth extension, the extension of the spectral envelope is of higher perceptual importance than that of the excitation signal [33].

Interpolation

The first step in any high bandwidth extension scheme is the resampling of the original signal such that the new sampling rate is sufficiently high for the representation of the extended signal. In Fig. 2.3, the original sampling rate is 8 kHz, however after interpolation, the rate is extended to 16 kHz. It is important to note that the actual frequency content of the signal is not modified at this point since the signal is still bandlimited at 4 kHz.

AR Coefficient Estimation

The estimation of the AR coefficients requires the use of a predictor that attempts to generate the wideband AR coefficients based upon certain features extracted from the narrowband signal and some *a priori* information available to the designer. For each frame of audio, a feature vector, $\mathbf{x}$, is calculated and fed to a pre-trained statistical model. The output of this statistical model is used in conjunction with already available information on speech production in order to estimate the AR coefficients of the wideband speech signal.

Analysis Filter

The analysis filter generates the narrowband excitation, $u_{nb}(t)$, signal from the narrowband speech signal and from the narrowband AR coefficients, $\mathbf{a}_{nb}$. This filter and the corresponding residual are defined in (2.1) and (2.2):

$$A_{nb}(z) = \sum_{i=0}^{P} a_{nb}(i) z^{-i} \tag{2.1}$$

$$u_{nb}(t) = \sum_{i=0}^{P} a_{nb}(i) s_{nb}(t - i) , \tag{2.2}$$

where P is the order of the AR model.

Extension of the Excitation

A number of methods exist for the extension of the excitation signal [55–57]. In general, these methods extract model parameters from the narrowband excitation and base the extension on these parameters. The model parameters are σ (the energy of the excitation), V (voiced/unvoiced switch), and F_0 (the fundamental frequency). A number of different techniques for wideband excitation generation are presented in greater detail in the ensuing section.

Synthesis Filter

Once the estimate of the wideband excitation signal, $u_{wb}(t)$, and the estimate of the wideband AR coefficients, $\mathbf{a}_{wb}$, is complete, the synthesis filter can use both constituents to form the speech estimate in the missing band. This is then filtered and added to the narrowband speech to form the final wideband speech signal shown in (2.3):

$$s_{wb}(t) = [h_{wb}(t) * u_{wb}(t)] * f(t) + s_{nb}(t) , \qquad (2.3)$$

where $h_{wb}(t)$ is the impulse response of $\frac{1}{A_{wb}(z)}$ and $f(t)$ is the impulse response of the high-pass filter that restricts the generated speech to the missing band.

The five steps described above are present in most source/filter-based wideband recovery algorithms. The proposed algorithms typically differ in the methods used to estimate the wideband envelope and the extension of the excitation. In the ensuing sections we describe some existing methods for envelope estimation and for wideband excitation signal generation.

2.2.1 ENVELOPE ESTIMATION

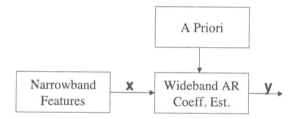

Figure 2.4: A general framework for wideband envelope estimation.

In the previous section we discussed a general framework for performing wideband recovery based on the source/filter model. Although the perceptual quality of the synthesized audio depends upon both the extended excitation and the estimated envelope, it has been shown that the estimation of the envelope is especially critical in wideband recovery algorithms [32]. As shown in Fig. 2.4, frequency regeneration algorithms use certain narrowband features and *a priori* knowledge in order to estimate the wideband spectral envelope. This estimation is typically performed using pre-trained statistical models, codebook mappings, or linear transformation algorithms on a frame-by-frame basis.

Although the general framework for envelope extension (Fig. 2.4) has been widely accepted, it is still unclear what method for performing the extension is optimal. In this section we focus on four popular methods for wideband envelope reconstruction. The first three follow the framework depicted in Fig. 2.4, whereas the fourth deviates from this framework by sending side information to encode the high-band envelope. These methods include:

- codebook mappings;

- linear transformations;

- statistical prediction methods;

- transmitted side information.

Codebook Mapping

The most commonly used method for estimating wideband envelopes is the codebook mapping approach. The underlying assumption to this method is that only a limited set of envelope observations occur. In other words, the envelopes of the high-band come from a typical set of speech sounds. For envelope reconstruction, no additional information is transmitted in the codebook formulation, but rather a pair of coupled codebooks are used. The first codebook is formed from a training set of narrowband feature vectors, whereas the second codebook is formed from corresponding wideband envelope representation vectors [33, 34].

A diagram of the algorithm is depicted in Fig. 2.5. Each incoming frame is analyzed and a narrowband feature vector, $\mathbf{x}$, is extracted. The narrowband codebook entry closest to the acquired feature vector is selected. In parallel to the narrowband codebook, there exists a wideband envelope "shadow" codebook that contains the wideband envelope associated with the selected narrowband codebook entry. The wideband envelope estimate, $\hat{\mathbf{y}}$, is therefore the entry of the wideband codebook coupled to the selected narrowband code vector.

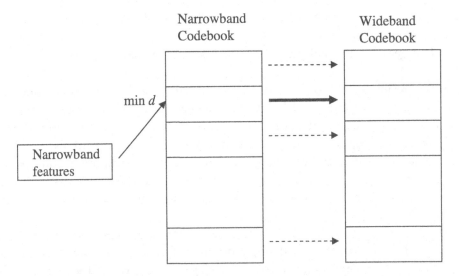

Figure 2.5: The narrowband and the wideband (shadow) codebooks. The narrowband codevector minimizing a distortion d between the extracted features and itself is found. The corresponding wideband parameters in the shadow codebook are selected for the envelope representation.[32]

It is clear from the above formulation that all estimated envelopes come from a limited set of envelopes obtained through training. The benefit of this approach is the guarantee that the resulting high-band LP filter will be stable. This restriction is also limiting, however, since the quality is restricted by the number and quality of the code vectors. To improve this, a number of papers have proposed envelope estimation based on a linear combination of codebook entries (Fig. 2.6) as shown in (2.4). This can be described as

$$\hat{\mathbf{y}} = \sum_{i=1}^{N_s} w_i \hat{\mathbf{y}}_i \, , \tag{2.4}$$

where w_i represents the weight that is given to wideband code vector $\hat{\mathbf{y}}_i$ and $\sum_{i=1}^{N_s} w_i = 1$. Although the weights, w_i, can be obtained using any number of methods, they are typically inversely proportional to the distortion distances between the acquired feature vector, $\mathbf{x}$, and narrowband codebook entry i.

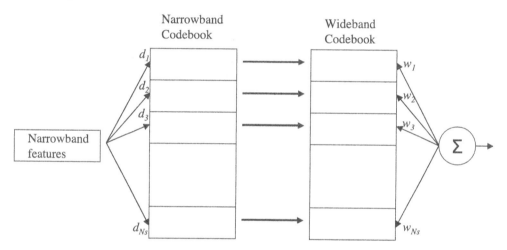

Figure 2.6: An alternate implementation of the codebook mapping algorithm. The distortions d_i between the extracted features and all other codevectors is found. The output is a linear combination of all codevectors with weights, w_i, inversely proportional to the distortions d_i.[32]

The Primary Codebook

The primary narrowband codebook is defined through vector quantization of the extracted narrowband features denoted by $\mathbf{x}$. In this section, we provide a high level overview of the training procedure for the narrowband codebook. The specific details of vector quantization in general can be found in [58].

A vector quantizer, Q, maps the b-dimensional feature space to a finite subspace : $Q :$ $\mathbf{R}^b \rightarrow C_{\mathbf{x}}$. The subspace is defined by set of vectors $C = \mathbf{x}_i : i = 1, 2, \ldots, N_s$. A number of different criteria for defining Q have been proposed, however most are based on the minimization of some distortion measure. Mathematically, this can be expressed as follows:

$$Q(\mathbf{x}) = \operatorname*{argmin}_{\hat{\mathbf{x}}_i \in C_x} d(\mathbf{x}, \hat{\mathbf{x}}_i) \qquad (2.5)$$

In order to adequately represent a data set, the length of the codebook can be quite large. One of the drawbacks of long codebooks is the extensive search required when mapping a new vector. Computationally efficient, but sub-optimal, schemes have been developed. As an example, in speech coding specifically, the development of the easy to search algebraic codebook for the quantization of the excitation is considered one of the breakthroughs in speech compression [59].

Using the mapping in (2.5), the quantizer assigns a Voronoi region to each code vector. This region is defined by:

$$V_i = \{\mathbf{x} \in \mathbf{R}^b : d(\mathbf{x}, \hat{\mathbf{x}}_i) \leq d(\mathbf{x}, \hat{\mathbf{x}}_j), \ \forall \ j \neq i\}. \qquad (2.6)$$

The Euclidean space, $\mathbf{R}^b$, is spanned by the union of the non-overlapping Voronoi regions that define the quantizer. In other words, $\bigcup_{i=1}^{N_s} V_i = \mathbf{R}^b$ and $V_i \bigcap V_j = \emptyset$ for any $j \neq i$.

Given a large set of training vectors $\mathbf{x}(m), m = 0 \ldots N_m - 1$, the objective becomes to minimize the mean quantization distortion. Mathematically, the codevectors are modified such that the functional in (2.7) is minimized with respect to $\hat{\mathbf{x}}_i$'s :

$$\frac{1}{N_m} \sum_{m=0}^{N_m-1} \min_{\hat{\mathbf{x}}_i \in C_x} d(\mathbf{x}(m), \hat{\mathbf{x}}_i) \qquad (2.7)$$

The LBG algorithm, a variant of the generalized Lloyd algorithm, is most often used for the training of the codebook [58]. The algorithm effectively forms a clustering of the data, with the centroids representing the codevectors.

The Shadow Codebook

The performance of the codebook mapping technique greatly depends upon the coupling between the narrowband and the wideband (shadow) codebook. Let $C_y = \{\hat{\mathbf{y}}_i \in \mathbf{R}^n | i = 1 \ldots N_s\}$ and let $F : \mathbf{R}^n \to C_y$. The codevector from the shadow codebook is selected as follows:

$$F(\mathbf{x}) = \hat{\mathbf{y}}_i \quad \text{where} \quad i = \operatorname{argmin} d(\mathbf{x}, \hat{\mathbf{x}}_i) \qquad (2.8)$$

In other words, the wideband envelope coupled with the narrowband feature vector that minimizes $d(\mathbf{x}, \hat{\mathbf{x}}_i)$ is selected. Due to the fixed relationship between the narrowband and wideband codebooks, it is clear from the above formulation that the narrowband codebook must first be trained before we can determine the appropriate wideband mapping function, $F(\mathbf{x})$ (or equivalently, the entries in C_y). The entries of the wideband codebook are then determined by clustering the data using the narrowband codebook, C_x. This procedure is shown in Fig. 2.7. As the diagram shows, the narrowband codebook is first generated. The wideband codevector of index i is formed from the centroid of all highband training envelopes coupled to the narrowband training vectors that lie in Voronoi region V_i [60].

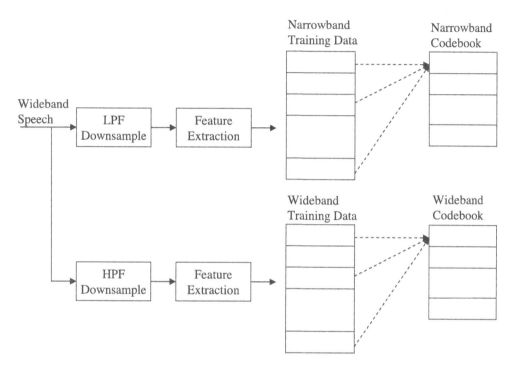

Figure 2.7: The coupling between the narrowband and the wideband (shadow) codebooks. After training the narrowband codebook, the wideband envelopes of the training data regions belonging to each centroid of the narrowband codebook are combined to form the wideband envelope parameters.[32]

The performance of envelope extension techniques based on vector quantization greatly depend on a number of factors, such as the choice of representation of the envelopes, $\mathbf{x}$ and $\mathbf{y}$, the selected distortion measure, $d(\mathbf{x}, \hat{\mathbf{x}})$, and the size of the codebook. Most often, $\mathbf{x}$ and $\mathbf{y}$ are some representation of the narrowband and wideband envelopes (e.g., AR coefficients, cepstral coefficients, reflection coefficients). As such, the distortion measure used in other speech coding algorithms are often used (e.g., log spectral distortion, cepstral distance). Although the distortion measures for the primary and shadow codebooks need not be the same, most implementations in the literature maintain the same distortion measures for both [33, 39].

Linear Mapping

Another set of methods for performing wideband envelope extension rely on a linear mapping from a narrowband feature space to a wideband envelope representation. Consider a set of extracted narrowband features represented by $\mathbf{x} = [x_1, x_2, \ldots x_p]$ and a vector of high-band envelope parameters $\mathbf{y} = [y_1, y_2, \ldots y_d]$. Mathematically, the wideband extension is performed as follows:

$$\tilde{\mathbf{y}} = \mathbf{A}^T \mathbf{x} , \tag{2.9}$$

where $\tilde{\mathbf{y}}$ is the wideband estimate (d-dimensional), $\mathbf{x}$ represents the extracted narrowband features (p-dimensional), and $\mathbf{A}$ is a $p \times d$ matrix representing a linear mapping learned through some training procedure.

Obtaining an appropriate matrix $\mathbf{A}$ requires training on an available wideband speech database. The speech is processed on a frame-by-frame basis. Narrowband features and the corresponding wideband envelope parameters are extracted from each frame and arranged in two matrices, as shown below:

$$\mathbf{F}_x = \begin{bmatrix} \mathbf{x}(0) \\ \mathbf{x}(1) \\ \vdots \\ \mathbf{x}(N-1) \end{bmatrix} \quad \mathbf{F}_y = \begin{bmatrix} \mathbf{y}(0) \\ \mathbf{y}(1) \\ \vdots \\ \mathbf{y}(N-1) \end{bmatrix} , \tag{2.10}$$

where $\mathbf{x}(i)$ and $\mathbf{y}(i)$ denote individual training vectors and N denotes the number of available speech segments in the training set.

The optimal matrix A should minimize:

$$e = \|\mathbf{y} - \mathbf{A}^T \mathbf{x}\|^2 \tag{2.11}$$

over the training set. The least squares approach of solving for $\mathbf{A}$ leads to the following minimization:

$$\epsilon^2 = tr\left[(\mathbf{F}_y - \mathbf{F}_x \mathbf{A})^T (\mathbf{F}_y - \mathbf{F}_x \mathbf{A}) \right] . \tag{2.12}$$

Taking a derivative with respect to each element of $\mathbf{A}$ and setting it to zero, we can solve for the optimal $\mathbf{A}$. This is given by:

$$\mathbf{A} = (\mathbf{F}_x^T \mathbf{F}_x)^{-1} \mathbf{F}_x \mathbf{F}_y . \tag{2.13}$$

Examples in the literature of linear mapping algorithms include [38] and [61]. In [38], Avendano *et al*. propose an algorithm that learns a linear mapping using as features the current frame envelope, the previous frame envelope, and the following frame envelope. In [61], Chennoukh *et al*. propose an algorithm that learns a linear mapping between narrowband LSF coefficients and high-band LSF coefficients.

Although linear mapping algorithms typically have low memory requirements and are easily implemented, there are two significant drawbacks associated with them. The first is that forcing a linear mapping between the low-band and high-band envelopes is restrictive. In addition, it is difficult to place restrictions on the estimated envelope parameters such that they are admissible. For example, through such mappings it is often the case that the estimated envelope parameters represent an unstable LP synthesis filter. Such restrictions have to be implemented through ad-hoc and heuristic methods. In [61] Chennoukh *et al.* derive an algorithm that maps narrowband LSF coefficients to wideband coefficients. Using this algorithm, the estimated wideband parameters may be greater than π. This is inadmissable therefore there are provisions in the algorithm that scale the parameters such that they fall within the admissible range.

Piecewise-Linear Mapping
The constraint that the mapping between the narrowband feature space and the high-band envelope parameter space must be linear is quite severe and it does not hold in general. A more promising approach to the problem is to partition the narrowband feature space into disjoint sets and train a linear mapping for each set [37, 61]. Assume that the feature space is divided into N_s sets by means of a vector quantizer. A mapping matrix $\mathbf{A}_i$ is trained for each partition i ($i = 1 \ldots N_s$) using the least-squares approach discussed above. Mathematically, each $\mathbf{A}_i$ is given by:

$$\mathbf{A}_i = (\mathbf{F}_{x,i}^T \mathbf{F}_{x,i})^{-1} \mathbf{F}_{x,i} \mathbf{F}_{y,i} \, . \tag{2.14}$$

Narrowband features outside the training set are first mapped to one of the N_s entries in the codebook generated by the VQ. This mapping determines which mapping matrix $\mathbf{A}_i$ is to be used for the feature set. This is depicted in Fig. 2.8.

Although the piecewise-linear mapping approach has been shown to improve the quality of the envelope estimate, the storage requirements are somewhat limiting. For a codebook of size N_s, N_s matrices of size $p \times d$ must be stored in addition to the classification codebook itself.

Other linear methods
In addition to the linear and piecewise linear methods discussed above, a number of other envelope extension methods exist that assume a linear relationship between the low band and the high band. In [39], the authors assume that the slope of the straight-line approximation of the narrowband log-scale spectrum is sufficient for shaping the generated high band spectrum. In Fig. 2.9, we show the actual spectrum of the narrowband spectrum and the corresponding straight-line approximation. This line is extended to shape the generated wideband signal and it also indirectly determines the gain of the high band. The efficacy of this technique depends upon the actual slope of the straight-line approximation for the narrowband and for the high band. For frames in which this slope is the same for both bands, the technique may work, however due to its simplicity, it often fails to capture the actual high-band spectrum.

Another linear technique [57] is based on a flat approximation of the high-band envelope. In this technique, a flat high-band is assumed when shaping the extended envelope. The method has proven to be successful for very limited high-band generation. For example, the technique is

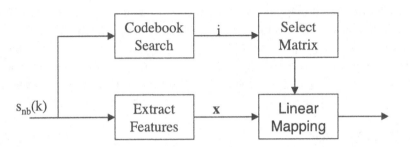

Figure 2.8: A diagram of the piecewise linear mapping method. The acquired feature vector is used to first classify the speech frame and the corresponding linear mapping matrix is selected.

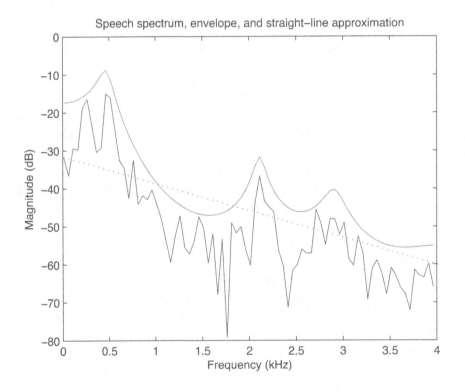

Figure 2.9: The magnitude spectrum of a voiced frame, the corresponding envelope, and the straight-line approximation.

used in the ITU G.722.2 standard adaptive-multi-rate wideband (AMR-WB) algorithm for the generation of the band from 6.4–7 kHz [57]. .

Additional techniques that rely on fixed envelopes have also been proposed in the literature. A common method for determining the fixed envelopes is to perform averages of high-band envelopes from a large database of voiced and unvoiced frames. In [30], a shaping filter for voiced frames is generated based on the idea of averaging over multiple frames and this filter is applied to the generated wideband spectrum. In addition, in [54] each frame is further classified as voiced, sibilant, or stop consonant and an appropriate envelope is applied depending on the classification. The envelopes are generated through averaging over training data.

Statistical Methods for Envelope Estimation

The limitations of the codebook method and the linear mapping method for envelope estimation can be severe. In envelope estimation through codebook mapping, the high-band envelope is restricted to a small, finite set of possibilities (the codebook size). The performance can be improved by increasing the codebook size or through a fuzzy approach [39], however the limitation still exists. The linear model associated with the linear mapping techniques tends to be too simplistic as a wideband recovery function. As a result, the mappings often significantly deviate from the actual envelope.

A set of methods that attempts to alleviate these problems is based on statistical recovery of the underlying envelope. These methods rely on a nonlinear wideband recovery function that is based on the statistics of the underlying data. Using this method, the estimated envelope is not restricted to a finite set and the mapping function is not necessarily linear.

A representative technique for statistical recovery methods in wideband speech prediction is presented in [35]. The underlying assumption in this technique is that the narrowband and wideband envelopes stem from linear combinations of narrowband and wideband sources. Consider a narrowband speech segment of length K represented by $\mathbf{x} = [x_0, x_1, \ldots, x_K]$ and the corresponding high-band segment represented by $\mathbf{y} = [y_0, y_1, \ldots, y_K]$. We assume that $\mathbf{x}$ is generated by N random sources ($\lambda_i, i = 1 \ldots N$) and $\mathbf{y}$ is generated by M random sources ($\theta_j, j = 1 \ldots M$). Let $\alpha_{i,j} = p(\theta_j | \lambda_i)$ denote the probability that a source θ_j contributes to the highband speech given that λ_i contributes to the narrowband speech. Given the parameter set $A = \{\alpha_{i,j}\}$, $\Lambda = \{\lambda_i\}$, and $\Theta = \{\theta_j\}$, it is possible to construct a wideband recovery function $f(\mathbf{x}, A, \Lambda, \Theta)$.

Given a set of training sequences ordered in matrices $\mathbf{X}$ and $\mathbf{Y}$, and given a set of parameters $P = \{A, \Lambda, \Theta\}$, the joint distribution can be written as follows:

$$p(\mathbf{X}, \mathbf{Y}) = \prod_t \sum_{i=1}^{N} \sum_{j=1}^{M} p(\mathbf{x}_t, \mathbf{y}_t) . \qquad (2.15)$$

The parameter set, $P = \{A, \Lambda, \Theta\}$, is estimated by maximizing the likelihood, $p(\mathbf{X}, \mathbf{Y}, P)$ using the expectation maximization algorithm (EM). During the E step, the log-likelihood is computed, and during the M step, the log-likelihood is maximized by varying P. Update equations for the parameters are provided in [35].

Once the parameters are updated, the MMSE estimate of the missing high band can be obtained using Equations (2.16)–(2.18):

$$\hat{\mathbf{Y}} = E[\mathbf{Y}|\mathbf{X}] = \int \mathbf{Y} \prod_t p(\mathbf{y}_t|\mathbf{x}_t)d\mathbf{Y} \tag{2.16}$$

$$p(\mathbf{y}_t|\mathbf{x}_t) = \sum_{i=1}^{N}\sum_{j=1}^{M} \frac{p(\mathbf{y}_t, \mathbf{x}_t, \lambda_i, \theta_j)}{p(\mathbf{x}_t)} \tag{2.17}$$

$$= \sum_{i=1}^{N}\sum_{j=1}^{M} \frac{p(\mathbf{y}_t|\theta_j)\alpha_{i,j}\,p(\mathbf{x}_t|\lambda_i)p(\lambda_i)}{\sum_{i=1}^{N} p(\mathbf{x}_t|\lambda_i)p(\lambda_i)}. \tag{2.18}$$

In addition to the statistical recovery function presented in [35], the authors in [11, 12, 21, 36] also present a statistical method based on a Gaussian Mixture Model (GMM).

Envelope Generation Using Transmitted Side Information
Bandwidth extension has also been used in the context of speech compression. Most speech compression schemes relying on bandwidth extension simply encode the high-band spectral envelope as either Line Spectral Frequencies (LSF) or cepstral coefficients and artificially generate the high-band excitation [9, 19, 20]. One of the more promising methods in this group is the newly standardized G.729.1 codec [19]. A split-band technique is used in this standard to encode the speech. At the encoder, the wideband speech is split into a high band and low band. The low band is encoded using the existing G.729 narrowband codec, and the high band is coarsely parameterized. In Fig. 2.10 we show a block diagram of the G.729.1 high-band encoder/decoder structure. In this algorithm, a frequency domain and a time domain envelope are extracted and quantized at the encoder from the high-band signal, $s_{HB}(t)$. At the decoder, the decoded envelopes are used to shape an artificially generated excitation signal and the high-band speech is formed.

At the encoder, the time domain envelope, $T_{env}(i)$, of $s_{HB}(t)$ is constructed as follows:

$$T_{env}(i) = \frac{1}{2}\log_2\left(\sum_{n=0}^{9} s_{hb}^2(n + 10i)\right), i = 0, ..., 15. \tag{2.19}$$

The resulting 16-point envelope is mean-removed split vector quantized. The mean of the envelope, M is encoded using 5 bits using 3 dB steps in the log domain. The mean value is then subtracted and the envelope is split in two 8-element vectors, each of which is vector quantized.

The frequency domain envelope, $F_{env}(i)$ is composed of the energy values of 12 equally spaced subbands in the high band. To generate the envelope, the high-band signal, $s_{HB}(t)$, is windowed by a Hanning window, and an FFT of the resulting signal is taken. The envelope is

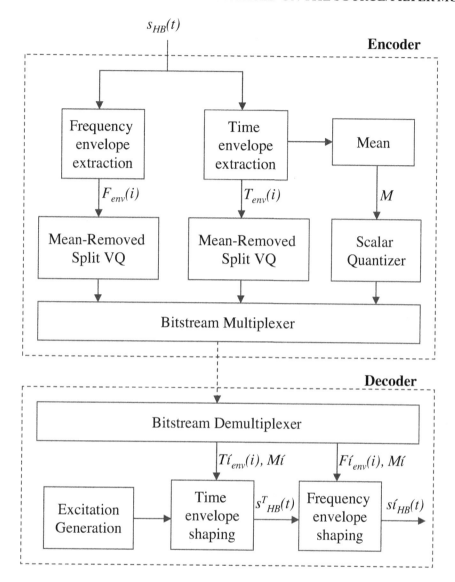

Figure 2.10: The high-band encoder and decoder of the G.729.1 speech compression standard.

generated by calculating the energy in the 12 equally spaced bands. The resulting envelope is split VQ quantized (using three codebooks) and embedded in the bitstream.

At the decoder, the artificially generated excitation, $u_{HB}(t)$, is scaled in time by an interpolated version of the decoded envelope to form $s_{HB}^{T}(t)$

$$s_{HB}^{T}(t) = g_T(t)u_{HB}(t) \,, \tag{2.20}$$

where $g_T(t)$ is obtained by interpolating the decoded time-domain envelope, $T'_{env}(i)$, using a "flat top" Hanning window. Following the time scaling, the spectrum of the resulting waveform is scaled to form the high-band speech. The scaling is done using a filterbank equalizer, where the gain of each filter corresponds to the decoded spectral envelope values. The impulse response of the equalizer is given by:

$$h_F(t) = \sum_{i=0}^{11} G_F(i)h_F^{(i)}(t) \,, \tag{2.21}$$

where $G_F(i)$ is a function of the decoded frequency envelope, $F'_{env}(i)$, and $h_F^{(i)}$ is the impulse response of the ith equalization filter as described in [19]. The high-band speech, $s'_{HB}(t)$, is obtained by filtering $s_{mb}^{T}(t)$ with the filter in 2.21.

Although these recent techniques that transmit additional information greatly improve speech quality when compared to techniques solely based on prediction, no explicit psychoacoustic models are employed for high-band synthesis. Hence, the bit rates associated with the high band representation are often unnecessarily high.

2.2.2 EXCITATION SIGNAL GENERATION

Most source/filter based frequency regeneration systems include a subsystem responsible for the extension of the excitation signal (see Fig. 2.3). This subsystem takes as an input the bandlimited excitation signal estimate, $u_{nb}(t)$, and produces a wideband estimate, $u_{wb}(t)$. The generated wideband excitation signal will serve as the input to the wideband synthesis filter, as shown in Fig. 2.3. Algorithms for performing the extension benefit from the simple periodic structure of the excitation and from the fact that the human ear is insensitive to high-band imperfections in reproduced audio. A number of methods for performing this extension are discussed in the literature [10, 19, 57, 62]. In this section we discuss in detail five popular methods for performing this extension:

- explicit generation of the high band;

- pitch scaling;

- spectral translation and folding;

- glottal pulse waveform extension;

- modulation of narrowband subbands.

We provide specifics for each method, discuss the benefits and drawbacks of each method, and provide some frequency domain plots of the extended excitations.

Explicit generation of the high band

Although a number of models exist for high-band excitation signal generation, perhaps the most straightforward is the generation of the excitation signal based on the source parameters. These parameters include: the fundamental frequency of the speech signal F_0, the voiced/unvoiced state V, and the gain factor, σ. Different approaches to the generation of the signal can be taken depending upon the allowed complexity in the overall system.

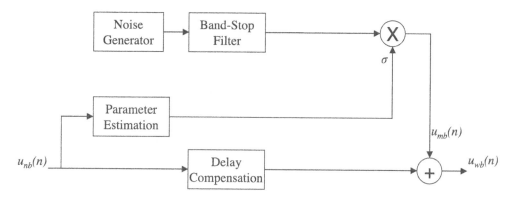

Figure 2.11: Excitation extension based on noise scaling.

A simple approach to high-band generation in the excitation signal is to simply add noise of appropriate energy in the high band (Fig. 2.11), disregarding the voiced/unvoiced state and the fundamental frequency. A noise generator forms the noise with gain factor σ, which is then filtered such that only the higher frequencies are present. The resulting signal, $u_{mb}(t)$, is then added to an appropriately delayed narrowband version of the signal and the wideband estimate of the excitation, $u_{wb}(k)$, is formed. Although this approach has been shown to underperform when regenerating large missing bands, it has been successful in narrowband frequency regeneration. For example, the adaptive multi-rate wide band (AMR-WB) algorithm uses this method for the generation of frequencies from 6.4–7 kHz [56, 57]. For extension from 4–7 kHz, however, the synthesized speech contains audible, annoying artifacts and the technique is not appropriate. By incorporating more narrowband parameters, we can obtain better results.

A natural refinement to the algorithm is the inclusion of the two disregarded parameters, namely the fundamental speech frequency and the voiced/unvoiced state variable. This is done by incorporating a voiced/unvoiced switch and a harmonics generator as shown in Fig. 2.12. The techniques for sine-wave generation are similar to those defined in [63] and [64]. The difference between the noise only system and this current system is that the current system discriminates between voiced and unvoiced frames. Previously, the excitation of all types of frames was extended using a filtered noise model, however the current model uses filtered noise during unvoiced frames and a tonal excitation during unvoiced frames. Although this technique performs better than the noise-only model, it is highly dependent on an accurate estimate of the fundamental frequency

F_0 for producing high quality speech. If the estimate of F_0 is incorrect, an impression is produced that an interfering simultaneous speaker whose pitch is slightly different is added to the original signal.

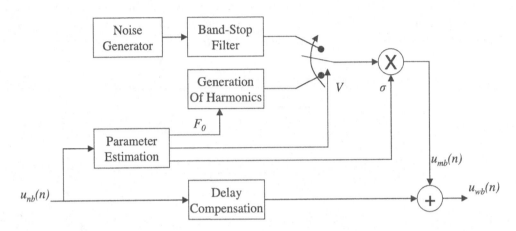

Figure 2.12: The block diagram of an excitation extension algorithm based on a noise+harmonics model.

To see the difference between a noise only model and one incorporating pitch information consider Fig. 2.13. In Fig. 2.13 (a) we show the actual high-band excitation (shifted in frequency for a narrowband representation). Fig. 2.13 (b) shows the extended excitation using a noise only model, whereas (c) shows the extended excitation using the noise+tone model. It is clear that the figure in part (c) better represents the harmonic structure of the actual excitation.

Pitch Scaling

One of the major disadvantages of the explicit generation technique is the fact that the parameters of the source model must be estimated. In an effort to remove this necessity, new methods for the extension of the excitation make use of pitch scaling. Figure 2.14 shows the general block diagram for the pitch scaling technique. The first step in the process is the downsampling of the narrowband excitation signal, $u_{nb}(t)$. The pitch of the resulting waveform is then doubled (time-stretching [65, 66]) and $u_{mb}(t)$ is formed. $u_{mb}(t)$ is high pass filtered and then added to a delayed version of the original signal. It is obvious from this technique that the spacing between the harmonics in the lower band is twice as short as it is in the high band (due to the pitch scaling by 2), however for most speakers this difference does not produce audible artifacts. The real advantage of this method is the fact that the estimates of the source parameters are not required, therefore the method tends to be quite robust [32].

In Fig. 2.15 we show the spectrum of the narrowband excitation (a) and the spectrum of the extended excitation (b). The pitch doubling is apparent in the extended excitation when

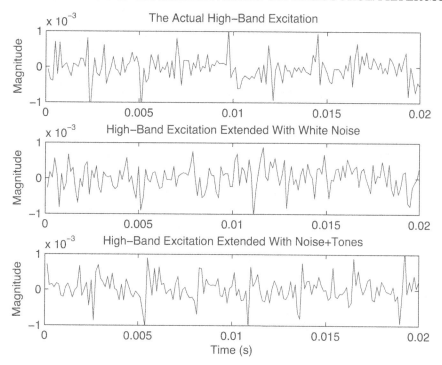

Figure 2.13: (a) The narrowband excitation and (b) the artificially extended high-band excitation using pitch scaling.

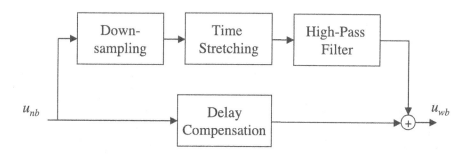

Figure 2.14: A block diagram of an excitation extension algorithm based on pitch scaling.

compared to the narrowband excitation. The peaks in frequency are twice as far apart in the extended excitation when compared to the original.

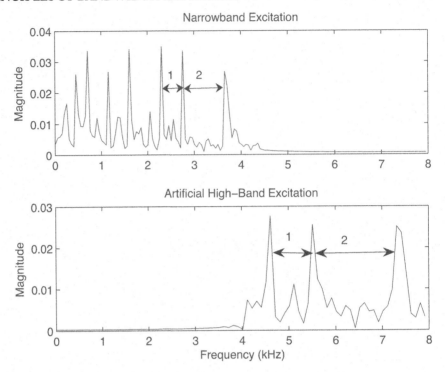

Figure 2.15: (a) The narrowband excitation and (b) artificially extended high-band excitation using pitch scaling.

Spectral Folding and Spectral Translation

The low-band and high-band arise from the same physical system therefore there is often a great deal of correlation between the narrowband excitation and the high-band excitation. This fact gives rise to techniques which simply use a frequency-translated or a frequency-folded version of the low-band excitation in the high band. A number of such algorithms have been proposed in the literature and most are based on the system described below [11, 27, 33, 67].

In spectral translation, the narrowband excitation signal is modulated, which causes a translation in frequency to the missing frequency bands. In the simplest scenario, the resulting signal goes through an anti-aliasing filter and is then is added to the narrowband excitation to form the wideband excitation. Mathematically, we can express the technique as follows:

$$u_{mb}(t) = [u_{nb}(t)cos(\Omega_m n)] * f(t) \tag{2.22}$$

$$u_{wb}(t) = u_{nb}(t) + u_{mb}(t) \tag{2.23}$$

where $f(t)$ is a filter that restricts the modulated narrowband excitation within the missing band.

The spectral folding technique is based on upsampling the narrowband excitation. Typically an upsampler would be followed by the low-pass filter to remove the duplicated high-band spectrum; however, in this case, we use the resulting signal as the final wideband excitation $u_{wb}(t)$.

Glottal Pulse Waveform Extension

In addition to extending the excitation, recently a number of techniques based on the extension of the glottal pulse waveforms have been proposed. In [62] the authors propose a technique that uses pitch-synchronous analysis. For voiced segments, the glottal closure instants are calculated from the narrowband excitation, whereas for unvoiced segments they are taken to be equally spaced. The method proposes changing the open quotient (OQ) associated with the narrowband glottal flow signal through a time scale transformation of the waveform. The open quotient is defined as follows:

$$OQ = \frac{T_a + T_e}{T},$$
(2.24)

where T_e is the duration of peak flow, T_a is the duration of the return phase, and T is the estimated pitch period. These values are labeled on the glottal flow waveform and its corresponding derivative in Fig. 2.16. This scaling extends the spectrum of the narrowband excitation using a

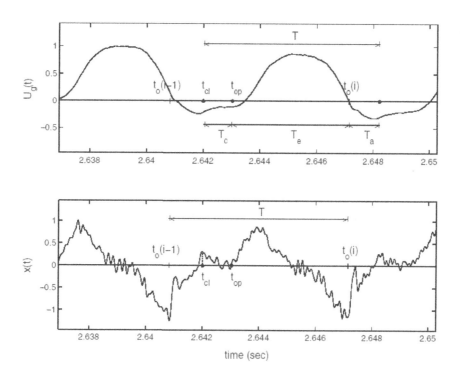

Figure 2.16: (a) The glottal waveform and (b) its derivative.[62]

similar approach to the ones used in the pitch scaling algorithms. The limitation of this model is the requirement of a robust pitch marking algorithm. This is especially problematic for low-SNR scenarios.

In addition to the glottal pulse waveform scaling, methods that artificially generate the glottal pulse waveform have recently gained popularity. In the ITU G.729.1 standard [19], the excitation is artificially extended using glottal pulse waveforms. A fractional pitch lag estimate is extracted from the narrowband excitation. The high-band excitation is artificially extended by placing glottal pulse waveforms at integer pitch lags. The fractional portion of the pitch lag is used to select the shape of the waveform from a lookup table [19].

Modulation of Narrowband Subbands

The frequency resolution of the human auditory system is limited for frequencies above 4 kHz. This implies that pitch periodicity above 4 kHz is perceived through the time-domain envelope. In [10], Unno and McCree propose a system in which the time domain envelope of the narrowband signal in the band between 3–4 kHz is modulated by white noise in the high band. This technique effectively applies the time-domain envelope of the band-limited excitation from 3–4 kHz to the high-band excitation. It ensures that the periodicity in the band limited signal from 3–4 kHz is maintained throughout the missing band.

This technique is shown in Fig. 2.17 in which we show the narrowband excitation, the band limited signal and its envelope, the resulting artificial high-band excitation, and the actual high-band excitation. The periodicity of the narrowband excitation in Fig. 2.17 (a) is quite apparent, however the high-band excitation in Fig. 2.17 (d) is not as strong. The generated high-band excitation maintains a similar periodicity to the 3–4 kHz band, which seems to match well to the actual excitation.

The algorithm proposed in [68] by Qian and Kabal relies on the fact that the harmonics in the latter part of the narrowband excitation are similar to the harmonics in the high-band excitation. The proposed algorithm extracts the band from 3–4 kHz using a bandpass filter centered at 3.5 kHz. Three modulators shift the extracted band to the three bands in the missing high band 4–5 kHz, 5–6 kHz, and 6–7 kHz . A high level overview of the algorithm is shown in Fig. 2.18. The limitation of both of these techniques is the underlying assumption that the excitation between the 3–4 kHz band and the high-band excitation are similar.

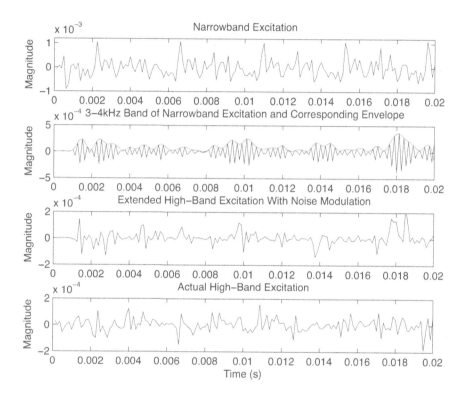

Figure 2.17: (a) The narrowband excitation, (b) the 3–4 kHz band of the narrowband excitation and the corresponding envelope, (c) the artificial high-band excitation, and (d) the true high-band excitation.

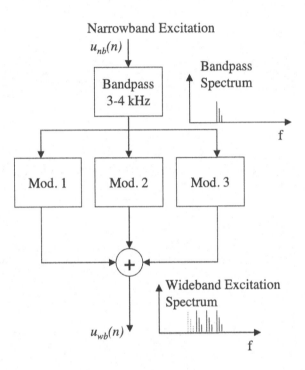

Figure 2.18: A block diagram of the technique in [68].

CHAPTER 3

Psychoacoustics

Today's audio compression algorithms rely upon models of the human ear to reduce the redundancies in the signal. Time and frequency properties of the ear are used to obtain a simplified model that enables a reduction in the information rate without affecting the perceptual quality of the audio. These models generally include the absolute threshold of hearing, critical band frequency analysis, and the masking properties of tones and noise. Audio coders combine these simplified psychoacoustic principles to shape the quantization noise such that it falls below audible thresholds. One example of a psychoacoustic model is the Perceptual Weighting Filter (PWF). The PWF places the quantization noise in areas of large signal energy so that it is masked by the signal. More thorough auditory models have also become parts of standardized audio compression algorithms, e.g., ISO/IEC MPEG-1 layer 3 [46], DTS [49], and Dolby AC-3 [50].

Although indirect psychoacoustic criteria are embedded in bandwidth extension algorithms (e.g., perceptual weighting filter) [43–45], there have been only a few attempts to include direct psychoacoustic principles [25, 52, 69]. In this section we give an overview of different psychoacoustic principles and their corresponding benefits and limitations.

3.1   GENERAL OVERVIEW

Researches have studied numerous mathematical representations of the human auditory system for the purpose of including them in audio compression algorithms. The most popular of these representations include the global masking threshold [70], auditory excitation pattern (AEP) [47], and perceptual loudness [48]. A masking threshold refers to a threshold below which a certain tone/noise signal is rendered inaudible due to the presence of another tone/noise masker. The global masking threshold is obtained by combining individual masking thresholds and it represents a threshold for a frame of audio, under which no frequency component is audible.

Auditory excitation patterns [47] describe the stimulation of the neural receptors caused by an audio signal. Each neural receptor is tuned into a specific frequency, therefore the AEP represents the output of each aural "filter" as a function of the center frequency of that filter.

Closely related to the AEP is the perceptual loudness [48, 71, 72]. Loudness is defined as the perceived intensity (in Sones) of the sound. The actual intensity (in dbSPL [1]) of a signal refers to an *external* measure of energy, whereas perceived loudness refers to an *internal* measure of

---

[1]All values of sound level are presented in terms of the Sound Pressure Level (SPL), a unit quantifying the intensity of a tone relative to an internationally defined reference of $m_0 = 20\mu Pa$. The intensity of a tone in SPL is defined as $I_{spl} = 20log(m/m_0)$, where m is the sound pressure of the stimulus in Pa [73].

energy. Because of the physiology of the human ear, the perceived internal energy is not equivalent to the actual energy of the signal. An example of two signals with equivalent energy but different levels of loudness is given in Section 3.2.2.

The psychoacoustic representations discussed above have been used in speech and audio compression algorithms, perceptual evaluation of speech and audio quality standards [74, 75], watermarking [76], speech enhancement [25], and gain control [77]. This section will discuss some of these principles in detail and provide a framework under which they can be used for bandwidth extension. A description of the human auditory system is first given, followed by a discussion on various masking phenomena. Afterwards, a brief review of existing psychoacoustic models is provided. Following this is a description of methods for computing auditory excitation patterns and perceptual loudness. Finally, a new bandwidth extension technique based on explicit psychoacoustic criteria is presented.

3.1.1   HUMAN AUDITORY SYSTEM

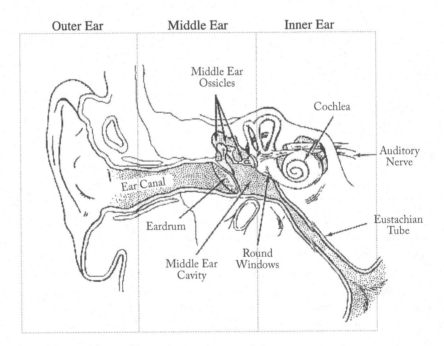

Figure 3.1: A physiological model of the human auditory system. The outer ear consists of the pinna and the ear canal. The middle ear consists of the ear drum and the middle ear ossicles. The inner ear contains the cochlea.[78]

A model of the human auditory system is shown in Fig. 3.1. The model consists of three different parts:

- the outer ear: the pinna and the ear canal;

- the middle ear: the ear drum and a mechanical transducer that converts the sound into mechanical vibrations;

- the inner ear: the cochlea, a spiral shaped bony-canal where the ear performs spectral analysis on the incoming sound. Converts vibrational energy produced by the middle ear into nerve impulses (i.e., transduction of mechanical energy to hydraulic energy to electrochemical energy).

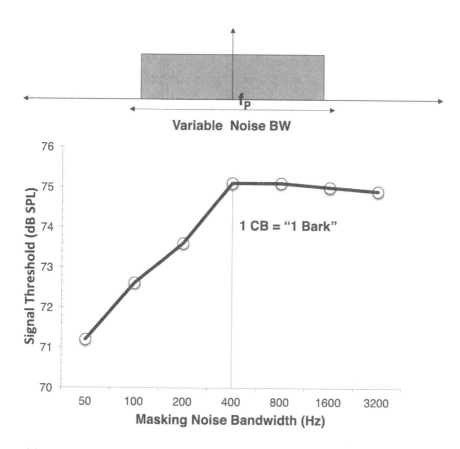

Figure 3.2: The results of Fletchers 1940 experiment. As the bandwidth of the noise source increases, so does the signal threshold, up until a certain frequency.[78]

In 1940, Fletcher [78] performed an important experiment that gave insight into the workings of the human auditory system shown above. His experiment consisted of a tone and wideband noise centered at the same frequency. Fletcher calculated the threshold of the tone as a function of the increasing noise bandwidth and his results are presented graphically in Fig. 3.2, which shows that as the bandwidth of the masker (wideband noise) is increased, the threshold of the tone is also increased, up until a certain point. After this point, the threshold of the signal does not vary with the changing bandwidth of the masker. Fletcher explained the results by suggesting that the auditory system contains an internal filter bank of bandpass filters. He states that the increasing noise bandwidth affects the threshold of the tone up until a certain point at which the auditory system filters the noise lying outside of its bandwidth.

Later experiments found that the filterbank discovered by Fletcher lies along the basilar membrane inside the cochlea, where a frequency to place transformation occurs. Different neural receptors positioned at different places along the basilar membrane, are tuned to different frequencies [48]. At places beyond these neural receptors, the sound waves decay rapidly and therefore a frequency to place transformation occurs. Figure 3.3 shows how a three tone stimulus is filtered along the basilar membrane. Specific neural receptors along the membrane are tuned to the three different frequencies and the excitation to receptors away from these points decays.

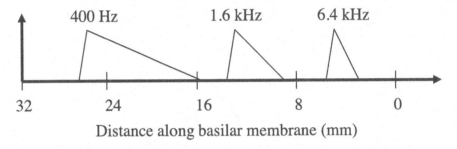

Figure 3.3: The neural receptors (lying along the basilar membrane) affected for the three tone-stimulus.

The filter passbands associated with the frequency to place transformation are highly overlapping and non-uniform [73, 78]. The term critical bandwidth refers to the bandwidths of these filters. A distance of 1 critical bandwidth is commonly referred to as "one Bark." The Bark scale [79], is defined as:

$$z_b(f) = 13 \, atan(0.00076f) + 3.5 \, atan\left[\left(\frac{f}{7500}\right)\right]. \tag{3.1}$$

The critical bandwidths are a function of the center frequency associated with that particular band and this relationship is shown in (3.2):

$$CB(f_c) = 25 + 75\left[1 + 1.4\left(\frac{f}{1000}\right)^2\right]^{0.69}. \tag{3.2}$$

Table 3.1 shows the idealized center frequencies and bandwidths of a collection of 25 critical bandwidth filters spanning the audible spectrum [79]. The bandwidths in this table have been obtained from the relationship in (3.2).

Table 3.1: Idealized center frequencies and bandwidths of a collection of 25 critical bandwidth filters spanning the audible spectrum [70, 79, 96]

Band	f_c (Hz)	BW (Hz)	Band	f_c	BW (Hz)	Band	f_c	BW (Hz)
1	50	...- 100	10	1175	1080-1270	19	4800	4400-5300
2	150	100-200	11	1370	1270-1480	20	5800	5300-6400
3	250	200-300	12	1600	1480-1720	21	7000	6400-7700
4	350	300-400	13	1850	1720-2000	22	8500	7700-9500
5	450	400-510	14	2150	2000-2320	23	10500	9500-12000
6	570	510-630	15	2500	2320-2700	24	13500	12000-15500
7	700	630-770	16	2900	2700-3150	25	19500	15500- ...
8	840	770-920	17	3400	3150-3700			
9	1000	920-1080	18	4000	3700-4400			

3.1.2 MASKING

Masking refers to the process where one sound (the maskee) is rendered inaudible due to the presence of another sound (the masker). Two types of masking are represented in modern audio compression algorithms, simultaneous masking and temporal masking. Simultaneous masking occurs if two sounds occur simultaneously and one is masked by the other, whereas temporal masking refers to the masking properties of signals occurring at slightly different points in time.

A listener detecting a tonal stimulus in background noise makes use of one of the auditory filters centered close to the frequency of the tone. This filter then not only resolves the tone, but it also removes noise at that center frequency with bandwidth greater than the critical bandwidth. The threshold of detection for the tonal signal is determined by the amount of noise present within that critical band. The first three examples below explain the concepts of simultaneous masking, whereas the last example depicts the effects of temporal masking.

No Masking

In this particular example, the tonal stimulus and the noise stimulus are located at different center frequencies, as shown in Fig. 3.4 (a). This means that no masking occurs since the stimuli lie in distant auditory filters, therefore the wideband noise stimulus has negligible effect on the threshold of the tonal stimulus.

Noise Masking Tone (NMT)

Under the NMT scenario, a noise stimulus is present at the same center frequency as a tonal stimulus and the tonal intensity is less than some threshold that depends upon the center frequency and the intensity of the masking noise. Figure 3.4 (b) shows the results of an experiment conducted in order to test the minimum signal to mask (SMR) ratio. As the figure indicates, the SMR is the difference between masking noise and the threshold of the masked tone. In this particular scenario, the noise masker is set at an intensity of 80 dB with a bandwidth of 1 bark and the threshold for detecting the tonal stimulus was determined to be 76 dB [79]. This corresponds to an SMR of 4 dB.

An interesting phenomena related to NMT is the fact that as the center frequency of the masker deviates from the center frequency of the tone, the SMR increases, meaning that the masking power of the noise stimulus decreases. This can be explained by once again examining the auditory filter bank. As the center frequency of the noise stimulus moves from the center frequency of the tone, part of the noise stimulus will fall in the stop band of the auditory filter, thereby decreasing the power of the noise masker.

Tone Masking Noise (TMN)

The TMN scenario is one in which a tonal stimulus is present at the same center frequency as a noise stimulus and the noise intensity is less than a threshold that depends upon the center frequency and the intensity of the masking tone. Figure 3.4 (c) shows the TMN scenario and also gives a typical SMR value of 24 dB [79]. Just as with NMT, as the center frequency of the masker tone deviates from the center frequency of the noise, the SMR increases thereby decreasing the power of the tone masker.

Simultaneous Masking

In addition to the simultaneous masking scenarios described above, temporal masking is also present. A weak sound that precedes/succeeds a louder sound is rendered inaudible. This is referred to as forward/backward temporal masking [79]. Figure 3.4 (d) gives an example of forward and backward temporal masking. A sound is played for 120 ms. For $\approx$ 20 ms prior to the onset of the stimulus and for $\approx$ 160 ms after the end of the stimulus, there is temporal masking that occurs within that critical band. The temporal masking curve is denoted by the dotted line.

3.2   EXISTING PSYCHOACOUSTIC MODELS

The masking effects of individual tones are not only present in the critical band in which the tones exist, but they spread to other bands as well. Most coding schemes represent this spread by means of a triangular individual masking threshold above the particular tone. The individual masking thresholds represent the maximum amount of noise that can be introduced at any frequency without changing the quality of the tone. Most audio frames, however, are much more spectrally complex than a simple tone. For these frames, the tonal components and their corresponding individual masking thresholds are identified within the frame. The individual masking thresholds of each tone are superimposed in each frame in order to form a Global Masking Threshold (GMT). The GMT gives insight to the amount of noise that can be introduced into the frame without creating any perceptual artifacts.

The ISO/IEC 11172-3 MPEG 1 standard has published two psychoacoustic models for determining the global masking threshold, psychoacoustic model 1, and psychoacoustic model 2 [46]. The general principle in both standards is to separate a spectrally complex signal into a sum

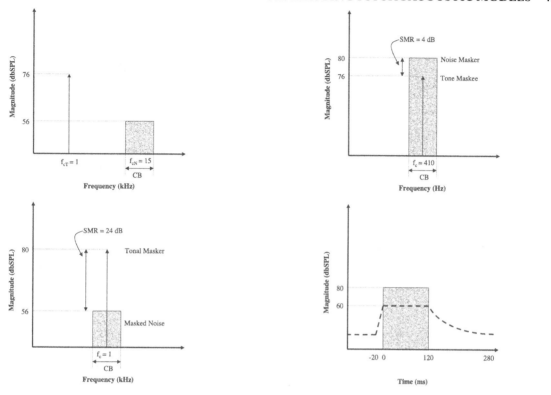

Figure 3.4: (a) No masking (b) Noise masking tone (NMT) (c) Tone masking noise (TMN) (d) Temporal masking. [70, 79]

of masker components (i.e., noise, tone) and compute a global masking threshold based on the masking properties of the individual components. Below we describe the steps in obtaining the GMT for both models, noting differences between the models when necessary.

Spectral estimation

The spectral magnitude of the frame is obtained by means of an FFT and the frame is normalized to the SPL scale. For psychoacoustic model 1, the window size is 512 samples, whereas for psychoacoustic model 2, the window size is 1024 samples.

Bark scale conversion

Both models analyze frequencies in groupings defined by the critical bandwidths previously discussed.

Tonal/non-tonal component estimation

Psychoacoustic model 1 identifies spectral peaks within each critical band as tonal components. The remaining components within each band are identified as a single non-tonal component. Psychoacoustic model 2 never explicitly calculates individual tonal and non-tonal components, but rather it computes a tonality index as a function of frequency. The tonality index is computed using a predictability measure. For each FFT bin, this index identifies the particular component as more tone-like or noise-like. Rather than making a binary decision (tone or noise), model 2 relies on intermediate tonality measures and determines a more exact masking threshold.

The spread of masking

Psychoacoustic model 1 determines the spread of intraband masking through an empirical model, whereas model 2 makes use of a spreading function. The resulting waveform is added to the threshold of hearing in a quiet environment and the global masking threshold is formed.

Just noticeable distortion

For each subband, a single masking threshold must be determined. Model 1 selects the minimum threshold within each band as the just noticeable distortion level for that band. Model 2 uses a combination of minimum thresholds and averages of the global masking threshold within each subband.

The just noticeable distortion curve contains a great deal of insight into the perceptually important and the perceptually redundant parts of the signal. All spectral components lying underneath the global masking threshold can be essentially removed without loss of signal quality. This is because any noise introduced by the removal of these spectral components is masked by the tonal/noise component closest to it. Furthermore, the real advantage of the GMT can be seen with respect to quantization. In areas of the spectrum where the tolerance for noise is higher (i.e., the SMR is low), more quantization noise can be introduced, and therefore less bits are required to encode those spectral components. This allows for a great deal of reduction in the overall bit rate compared to a PCM scheme.

Although the psychoacoustic models discussed in this section have been used in a number of standards [46, 49, 50], recent research suggests that perceptual methods based on excitation patterns may be more appropriate [51]. In the next section, we discuss the specifics of calculating excitation patterns.

3.2.1   AUDITORY EXCITATION PATTERNS

Auditory excitation patterns determine the amount of energy captured by each neural receptor for a particular aural stimulation. There are two prevalent methods for computing excitation patterns for a signal. The first method is based on an auditory filter model [73] and the second method is based on a parametric spreading function [80]. In this section, we briefly describe the first method.

The steps involved with the first method are shown in Figs. 3.5(a) and (b) for a 1 kHz tone. The first step in the method involves the computation of auditory filter shapes for a number of center frequencies (i.e., 750 Hz, 1000 Hz, 1300 Hz, 1700 Hz, 2000 Hz). The filter shape is computed using the model described in [73] as shown below:

$$W(g) = (1 + pg)e^{-pg} ,$$

(3.3)

where $g = \left| \frac{f_c - f}{f_c} \right|$ is the deviation from the center frequency, f_c. The shape of the filter is given by the parameter

$$p = \frac{4F_c}{24.7 \left[\frac{4.37f}{1000} + 1 \right]} .$$

(3.4)

Following the formulation of the auditory filter shapes, the points where the auditory filter intersects the 1 kHz stimulation are identified. The excitation pattern associated with this tone is obtained by plotting the points of intersection as a function of the center frequencies. An interpolation method (i.e., spline fit) can be used to interpolate in between the intersection points. This is shown in Fig. 3.5 (b). In this figure we plot the points labeled in Fig. 3.5 (a) and their interpolation. Although this method highlights the important concepts of excitation patterns, it is inexact and expensive to calculate. A more exact and computationally less demanding method for computing excitation patterns is described in the next section.

Parametric Spreading Function Approach
One of the problems associated with the method for computing AEP's discussed above, is the required computational complexity for obtaining high-resolution AEP's. A method introduced by Moore and Glasberg in the late 1980's is based on a parametric spreading function [80]. A high level overview is shown in Fig. 3.6. The figure shows the steps required for computing AEPs. The Hanning windowed spectrum of the signal of interest, $s(n)$, is first computed. The resulting spectrum is then normalized to an assumed playback level. Following the normalization, a middle and outer ear transfer function is applied to the normalized signal. The frequency response of the transfer function is shown in Fig. 3.7 [47].

The resulting spectral components are then converted from a linear scale to a Bark scale and then smeared according to a parametric spreading function as described in [80]. The smeared function represents the excitation pattern, E_p, corresponding to $s(n)$.

As an example to illustrate the method, consider a 1 kHz sinusoid sampled at 44 kHz. In Figs. 3.8 (a)–(d) we show the intermediate signals of the parametric spreading function approach. Fig. 3.8 (a) shows the original 2048 point signal. Fig 3.8 (b) shows the power spectral density after the middle and outer ear correction. Fig. 3.8 (c) shows the spreading function and Fig. 3.8 (d) shows the resulting AEP.

3.2.2   PERCEPTUAL LOUDNESS
This section provides details on the calculation of the loudness. Although a number of techniques exist for the calculation of the loudness, we describe the model proposed by Moore *et al.* [48].

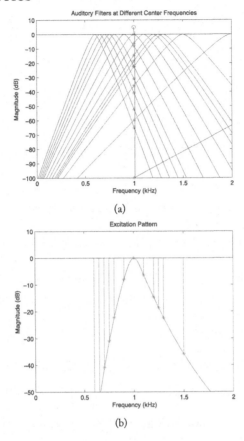

(a)

(b)

Figure 3.5: (a) Auditory filter shapes for different center frequencies and a tone at 1 kHz (b) and the AEP obtained by plotting the intersection point between the auditory filter shape and tone as a function of the center frequency.

Here, we give a general overview of the technique. A more detailed description is provided in the referred paper.

Perceptual loudness is defined as the area under a transformed version of the excitation pattern. A block diagram of the step-by-step procedure for computing the loudness is shown in Fig. 3.9. The excitation pattern (as a function of frequency) associated with the frame of audio being analyzed is first computed using the parametric spreading function approach [47]. In the model, the frequency scale of the excitation pattern is transformed to a scale that represents the human auditory system. More specifically, the scale relates frequency (F in kHz) to the number of equivalent rectangular bandwidth (ERB) auditory filters below that frequency [48]. The number

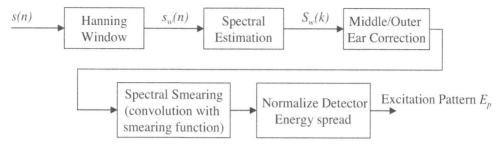

Figure 3.6: The block diagram of the parametric spreading function method for computing the auditory excitation pattern of a signal.

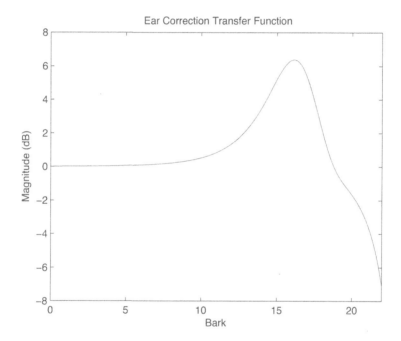

Figure 3.7: The transfer function of the middle and outer ear correction.

of ERB auditory filters, p, as a function of frequency, F, is given by:

$$p(F) = 21.4 \log_{10}(4.37F + 1). \tag{3.5}$$

As an example, for 16 kHz sampled audio, the total number of ERB auditory filters below 8 kHz is ≈ 33.

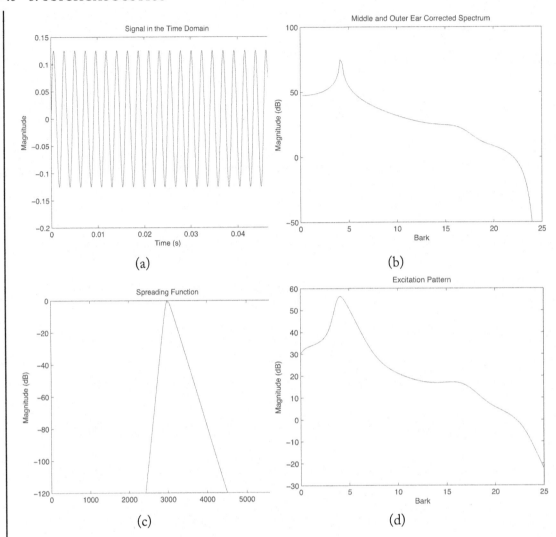

Figure 3.8: (a) Time domain plot of a 2048 point 1 kHz sinusoid sampled at 44 kHz, (b) the modified spectrum of the signal in part (a), (c) the spreading function, and (d) the resulting excitation pattern.

The specific loudness pattern as a function of the ERB filter number, $L_s(p)$, is next determined through a nonlinear transformation of the AEP as shown in:

$$L_s(p) = kE(p)^\alpha, \qquad (3.6)$$

where $E(p)$ is the excitation pattern at different ERB filter numbers, $k = 0.047$, and $\alpha = 0.3$ (empirically determined). Note that the above equation is a special case of a more general equation

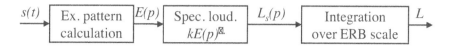

Figure 3.9: The block diagram of the method used to compute the perceptual loudness of each speech segment.

for loudness given in [48], $L_s(p) = k[(GE(p) + A)^\alpha - A^\alpha]$. The equation above can be obtained by disregarding the effects of low sound levels ($A = 0$), and by setting the gain associated with the cochlear amplifier at low-frequencies to one ($G = 1$). The total loudness can be determined by summing the loudness across the whole ERB scale, (3.7).

$$L = \int_0^P L_s(p)dp, \tag{3.7}$$

where $P \approx 33$ for 16 kHz sampled audio. Physiologically, this metric represents the total neural activity evoked by the particular sound.

In Fig. 3.10 we show the specific loudness patterns associated with two signals of equal energy. The first signal consists of a single tone (430 Hz) and the second signal consists of 3 tones (430 Hz, 860 Hz, 1720 Hz). Although the two signals have equivalent energy, the loudness associated with each is not equal. The loudness of the single-tone signal is 3.34 Sones, whereas the loudness of the 3-tone signal is 8.57 Sones. This example shows that although energy and loudness are correlated, they are not equivalent. In fact, it is easy to construct scenarios in which a signal has higher energy, but a lower loudness.

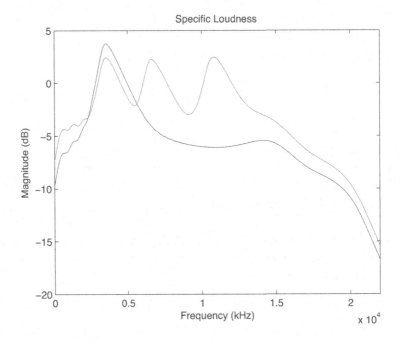

Figure 3.10: The specific loudness patterns of two signals with identical energy. The first signal consists of a single tone (430 Hz) and the second signal consists of 3 tones (430 Hz, 860 Hz, 1720 Hz). Although their energies are the same, the loudness of the single tone signal (12.1 Sones) is lower than the loudness of the 3-tone signal (15.2 Sones).

CHAPTER 4

Bandwidth Extension Using Spline Fitting

The public switched telephony network (PSTN) and most of today's cellular networks use speech coders operating with limited bandwidth (0.3–3.4 kHz), which in turn places a limit on the naturalness and intelligibility of speech [1]. This is most problematic for sounds whose energy is spread over the entire audible spectrum. For example, unvoiced sounds such as "s" and "f" are often difficult to discriminate with a narrowband representation. In Fig. 4.1, we provide a plot of the spectra of a voiced and an unvoiced segment up to 8 kHz. The energy of the unvoiced segment is spread throughout the spectrum, however most of the energy of the voiced segment lies at the low frequencies. The main goal of algorithms that aim to recover a wideband (0.3–7 kHz) speech signal from its narrowband (0.3–3.4 kHz) content is to enhance the intelligibility and the overall quality (pleasantness) of the audio. Many of these bandwidth extension algorithms make use of the correlation between the low band and the high band in order to predict the wideband speech signal from extracted narrowband features [10–13]. Recent studies, however, show that the mutual information between the narrowband and the high frequency bands is insufficient for wideband synthesis solely based on prediction [21–23]. In fact, Nilsson *et al.* show that the available narrowband information reduces uncertainty in the high band, on average, by only $\approx 10\%$ [23]. As a result, some side information must be transmitted to the decoder in order to accurately characterize the wideband speech. An open question, however, is how to minimize the amount of side information without affecting synthesized speech quality? In this chapter, we provide a possible solution through the development of an explicit psychoacoustic model that determines a set of perceptually relevant subbands within the high band. The sel

Most existing wideband recovery techniques are based on the source/filter model [10, 12, 13, 24]. These techniques typically include implicit psychoacoustic principles, such as perceptual weighting filters and dynamic bit allocation schemes in which lower frequency components are allotted a larger number of bits. Although some of these methods were shown to improve the quality of the coded audio, studies show that additional coding gain is possible through the integration of explicit psychoacoustic models [25–27, 81]. Existing psychoacoustic models are particularly useful in high-fidelity audio coding applications, however their potential has not been fully utilized in traditional speech compression algorithms or wideband recovery schemes.

In this chapter, we discuss a psychoacoustic model for bandwidth extension tasks. The signal is first divided into subbands. An elaborate loudness estimation model is used to predict how

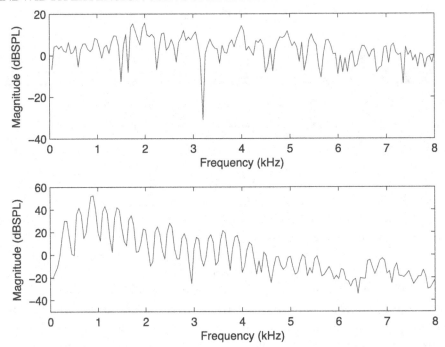

Figure 4.1: The energy distribution in frequency of an unvoiced frame (top) and of a voiced frame (bottom).

much a particular frame of audio will benefit from a more precise representation of the high band. A greedy algorithm is used to determine the importance of high-frequency subbands based on perceptual loudness measurements. The model is then used to select and quantize a subset of subbands within the high band, on a frame-by-frame basis, for the wideband recovery. A common method for performing subband ranking in existing audio coding applications is using energy-based metrics [19]. These methods are often inappropriate, however, because energy alone is not a sufficient predictor of perceptual importance. In fact, it is easy to construct scenarios in which a signal has a smaller energy, yet a larger perceived loudness when compared to another signal. One solution to his problem is to perform the ranking using an explicit loudness model proposed by Moore and Glasberg in [48].

In addition to the perceptual model, we also describe a coder/decoder structure in which the lower frequency band is encoded using an existing linear predictive coder, while the high-band generation is controlled using the perceptual model. The algorithm is developed such that it can be used as a "wrapper" around existing narrowband vocoders in order to improve performance without requiring changes to existing infrastructure. The underlying bandwidth extension algorithm is based on a source/filter model in which the high-band envelope and excitation are estimated separately. Depending upon the output of the subband ranking algorithm, the envelope is parameterized at the encoder, and the excitation is predicted from the narrowband excitation. We compare the proposed scheme to one of the modes of the narrowband adaptive multi-rate (AMR) coder and show that the proposed algorithm achieves improved audio quality at a lower average bit rate [82]. Furthermore, comparing this scheme to the wideband AMR coder shows comparable quality at a lower average bit rate [57].

4.1 OVERVIEW OF EXISTING WORK

In this section, we provide an overview of bandwidth extension algorithms and perceptual models. The specifics of the most important contributions in both cases are discussed along with a description of their respective limitations.

4.1.1 BANDWIDTH EXTENSION

Most bandwidth extension algorithms fall in one of two categories, bandwidth extension based on explicit high-band generation and bandwidth extension based on the source/filter model. Fig. 4.2 shows the block diagram for bandwidth extension algorithms involving band replication followed by spectral shaping [29] [30] [31]. Consider the narrowband signal $s_{nb}(t)$. To generate an artificial wideband representation, the signal is first upsampled,

$$\hat{s}_{1,wb}(t) = \begin{cases} s_{nb}(t/2) & \text{if } mod(t,2) = 0, \\ 0 & \text{else.} \end{cases} \tag{4.1}$$

This folds the low-band spectrum (0–4 kHz) onto the high band (4–8 kHz) and fills out the spectrum. Following the spectral folding, the high band is transformed by a shaping filter, $s(t)$,

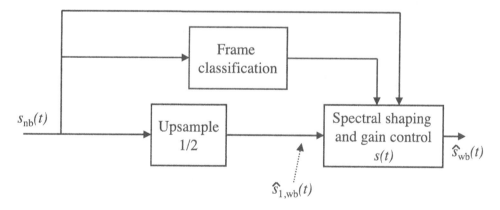

Figure 4.2: Bandwidth extension methods based on artificial band extension and spectral shaping.

$$\hat{s}_{wb}(t) = \hat{s}_{1,wb}(t) * s(t), \text{ where } * \text{ denotes convolution.} \tag{4.2}$$

Different shaping filters are typically used for different frame types. For example, the shaping associated with a voiced frame may introduce a pronounced spectral tilt, whereas the shaping of an unvoiced frame tends to maintain a flat spectrum. In addition to the high-band shaping, a gain control mechanism controls the gains of the low band and the high band such that their relative levels are suitable.

Examples of techniques based on similar principles include [29], [30], and [31]. Although these simple techniques can potentially improve the quality of the speech, audible artifacts are often induced. Therefore, more sophisticated techniques based on the source/filter model have been developed.

Most successful bandwidth extension algorithms are based on the source/filter speech production model [10] [11] [12] [13] [32]. The auto-regressive (AR) model for speech synthesis is given by:

$$\hat{s}_{nb}(t) = \hat{u}_{nb}(t) * \hat{h}_{nb}(t), \tag{4.3}$$

where $\hat{h}_{nb}(t)$ is the impulse response of the all-pole filter given by $\hat{H}_{nb}(z) = \frac{\sigma}{\hat{A}_{nb}(z)}$. $\hat{A}_{nb}(z)$ is a quantized version of the Nth order linear prediction (LP) filter given by:

$$A_{nb}(z) = 1 - \sum_{i=1}^{N} a_{i,nb} z^{-i}, \tag{4.4}$$

σ is a scalar gain factor, and $\hat{u}_{nb}(t)$ is a quantized version of

$$u_{nb}(t) = s_{nb}(t) - \sum_{i=1}^{N} a_{i,nb} s_{nb}(t-i). \tag{4.5}$$

A general procedure for performing wideband recovery based on the speech production model is given in Fig. 4.3 [32]. In general, a two step process is taken to recover the missing band. The first step involves the estimation of the wideband source-filter parameters, a_{wb}, given certain features extracted from the narrowband speech signal, $s_{nb}(t)$. The second step involves extending the narrowband excitation, $u_{nb}(t)$. The estimated parameters are then used to synthesize the wideband speech estimate. The resulting speech is high-pass filtered and added to a 16 kHz resampled version of the original narrowband speech, denoted by $s'_{nb}(t)$, given by:

$$\hat{s}_{wb}(t) = s'_{nb}(t) + \sigma g_{HPF}(t) * [h_{wb}(t) * u_{wb}(t)], \tag{4.6}$$

where $g_{HPF}(t)$ is the high pass filter that restricts the synthesized signal within the missing band prior to the addition with the original narrowband signal. This approach has been successful in a number of different algorithms [12, 32–38]. In [33] and [34], the authors make use of dual, coupled codebooks for parameter estimation. In [12, 35, 36], the authors use statistical recovery functions that are obtained from pre-trained Gaussian mixture models (GMM) in conjunction with hidden Markov models (HMM). Yet another set of techniques use linear wideband recovery functions [37, 38].

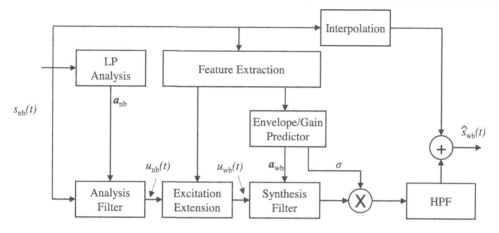

Figure 4.3: High-level diagram of traditional bandwidth extension techniques based on the source/filter model.

The underlying assumption for most of these approaches is that there is sufficient correlation or statistical dependency between the narrowband features and the wideband envelope to be predicted. While this is true for some frames, it has been shown that the assumption does not hold in general [21–23]. In Fig. 4.4, we show examples of two frames that illustrate this point. The figure shows two frames of wideband speech along with the true envelopes and predicted envelopes. The estimated envelope was predicted using a technique based on coupled, pre-trained codebooks; a technique representative of several modern envelope extension algorithms [39]. Figure 4.4 (top) shows a frame for which the predicted envelope matches the actual envelope quite well. In Fig. 4.4 (bottom), the estimated envelope greatly deviates from the actual and, in fact, erroneously introduces two high-band formants. In addition, it misses the two formants located between 4 and 6 kHz. As a result, a recent trend in bandwidth extension has been to transmit additional high-band information rather than using prediction models or codebooks to generate the missing bands.

Since the higher frequency bands are less sensitive to distortions (when compared to the lower frequencies), a coarse representation is often sufficient for a perceptually transparent representation [19, 40]. This idea is used in high-fidelity audio coding based on spectral band replication [40] and in the newly standardized G.729.1 speech coder [19]. Both of these methods employ an existing codec for the lower frequency band while the high band is coarsely parameterized using fewer parameters. Although these recent techniques greatly improve speech quality when compared to techniques solely based on prediction, no explicit psychoacoustic models are employed for high-band synthesis. Hence, the bit rates associated with the high band representation are often unnecessarily high.

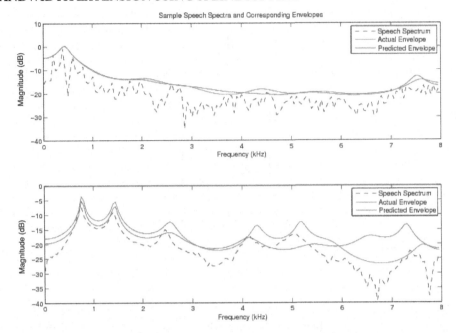

Figure 4.4: Wideband speech spectra (in dB) and their actual and predicted envelopes for two frames. The top figure shows a frame for which the predicted envelope matches the actual envelope. In the bottom figure, the estimated envelope greatly deviates from the actual.

4.1.2   PERCEPTUAL MODELS

Most existing wideband coding algorithms attempt to integrate indirect perceptual criteria to increase coding gain. Examples of such methods include perceptual weighting filters [43], perceptual LP techniques [44], and weighted LP techniques [45]. The perceptual weighting filter attempts to shape the quantization noise such that it falls in areas of high signal energy, however it is unsuitable for signals with a large spectral tilt (i.e., wideband speech). The perceptual LP technique filters the input speech signal with a filterbank that mimics the ear's critical band structure. The weighted LP technique manipulates the axis of the input signal such that the lower, perceptually more relevant, frequencies are given more weight. Although these methods improve the quality of the coded speech, additional gains are possible through the integration of an explicit psychoacoustic model.

Over the years, researchers have studied numerous explicit mathematical representations of the human auditory system for the purpose of including them in audio compression algorithms. The most popular of these representations include the global masking threshold [46], auditory excitation pattern (AEP) [47], and perceptual loudness [48].

A masking threshold refers to a threshold below which a certain tone/noise signal is rendered inaudible due to the presence of another tone/noise masker. The global masking threshold (GMT) is obtained by combining individual masking thresholds; it represents a spectral threshold that determines whether a frequency component is audible [46]. The GMT provides insight into the amount of noise that can be introduced into a frame without creating perceptual artifacts. For example in Fig. 4.5, at bark 5, approximately 40 dB of noise can be introduced without affecting the quality of the audio. Psychoacoustic models based on the global masking threshold have been used to shape the quantization noise in standardized audio compression algorithms, e.g., the ISO/IEC MPEG-1 layer 3 [46], DTS [49], and Dolby AC-3 [50]. In Fig. 4.5 we show a frame of audio along with its GMT. The masking threshold was calculated using the psychoacoustic model 1 described in the MPEG-1 algorithm [46].

Auditory excitation patterns (AEP) describe the stimulation of the neural receptors caused by an audio signal. Each neural receptor is tuned to a specific frequency, therefore the AEP represents the output of each aural "filter" as a function of the center frequency of that filter. As a result, two signals with similar excitation patterns tend to be perceptually similar. An excitation pattern-matching technique called excitation similarity weighting (ESW) was proposed by Painter and Spanias for scalable audio coding [51]. ESW was initially proposed in the context of sinusoidal modeling of audio. ESW ranks and selects the perceptually relevant sinusoids for scalable coding. The technique was then adapted for use in a perceptually motivated linear prediction algorithm [52].

A concept closely related to excitation patterns is perceptual loudness. Loudness is defined as the perceived intensity (in Sones) of an aural stimulation. It is obtained through a nonlinear transformation and integration of the excitation pattern [48]. Although it has found limited use in coding applications, a model for sinusoidal coding based on loudness was recently proposed [53]. In addition, a perceptual segmentation algorithm based on partial loudness was proposed in [51].

Although the models described above have proven very useful in high-fidelity audio compression schemes, they share a common limitation in the context of bandwidth extension. There exists no natural method for the explicit inclusion of these principles in wideband recovery schemes. In the ensuing section we propose a novel psychoacoustic model based on perceptual loudness that can be embedded in bandwidth extension algorithms.

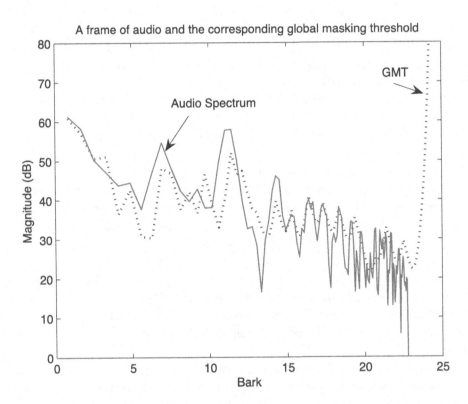

Figure 4.5: A frame of audio and the corresponding global masking threshold as determined by psychoacoustic model 1 in the MPEG-1 specification. The GMT provides insight into the amount of noise that can be introduced into a frame without creating perceptual artifacts. For example, at bark 5, approximately 40 dB of noise can be introduced without affecting the quality of the audio.

CHAPTER 5

Summary

In this book, the principles of speech bandwidth extension using perceptual criteria was discussed. Several methods and standard were discussed, along with applicable extensions. Bandwidth extension techniques were first introduced in Chapter 2. Explicit high-band generation was discussed as well as generation based on the source-filter model. Both envelope estimation and excitation signal extension were covered. In Chapter 3, psychoacoustics were discussed, including an overview of the human auditory system and a discussion on existing psychoacoustic models. Finally, Bandwidth extension using spline fitting was discussed in Chapter 4, along with its advantages and drawbacks.

This book serves as an introduction to speech bandwidth extension and provides a glimpse into some of its features, applications, and extensions. For further coverage interested readers are referred to the recent literature: [83, 84] use a method that avoids the introduction of inharmonicity and avoids the costly transmission of additional control parameters for frequency shifts; [85] uses a wideband speech coding scheme based on bandwidth extension and sparse linear prediction; [86] performs spectral band replication in the modified discrete cosine transform (MDCT) domain with no additional bits; [87] uses a hidden Markov model (HMM)-based wideband spectral envelope estimation by decoding an optimal Viterbi path based on the temporal contour of the narrowband spectral envelope and then performs the minimum mean square error (MMSE) estimation of the wideband spectral envelope on this path. In [88], binaural ABE is proposed. Most ABE methods have been developed for monaural signals and they cannot be applied as such to binaural signals due to a possible mismatch of binaural cues in the created frequency range; [89] uses context-dependent deep neural network hidden Markov model (CD-DNN-HMM) and Mel-scale log-filter bank features to achieve higher recognition accuracy than using MFCCs, but also formulate the mixed-bandwidth training problem as a missing feature problem, in which several feature dimensions have no value when narrowband speech is presented. This treatment makes training CD-DNN-HMMs with mixed-bandwidth data an easy task since no bandwidth extension is needed. [90] uses a neural network is used to estimate the mel spectrum in the extension band in short time frames based on features calculated from the narrowband speech; [91] extends the bandwidth of telephone speech to the frequency range 0–300 Hz. The method generates the lowest harmonics of voiced speech using sinusoidal synthesis. The energy in the extension band is estimated from spectral features using a Gaussian mixture model. In [92], speaker vocal tract shape information corresponding to the wideband signal is extracted by a codebook search. Postprocessing of the estimated vocal tract shape using iterative tuning allows artifacts reduction in cases of erroneous estimation of speech phoneme or vocal tract shape. In both [93, 94], a wideband line spectrum pain (LSP) codebook is coupled with the same index as the LSP codebook of a narrowband speech codec. The received narrowband LSP codebook indicies are used to directly induce wideband LSP codewords. Thus, the proposed scheme eliminates the codebook search processing to estimate the wideband spectrum envelope.

Notation

h	impulse response
g	gain control function
s	shaping filter
s_{nb}	narrowband shaping filter
s_{wb}	wideband shaping filter
u	excitation signal
u_{nb}	narrowband excitation signal
u_{wb}	wideband excitation signal
x	narrowband parameter
y	wideband parameter
$\mathbf{a}$	AR coefficients
$\mathbf{a}_{nb}$	narrowband AR coefficients
$\mathbf{a}_{wb}$	wideband AR coefficients
A	AR filter
Q	vector quantizer
T_{env}	time envelope
F_{env}	frequency envelope
F_0	fundamental frequency
V	voiced/unvoiced indicator
σ	excitation energy
OQ	open quotient
T	pitch period
T_a	duration of peak flow
T_e	duration of return phase

Bibliography

[1] A. Spanias, "Speech coding: A tutorial review," in *Proc. of IEEE*, vol. 82, no. 10, October 1994. DOI: 10.1109/5.326413. 1, 49

[2] N. Benveuto, "The 32 kbit/s adpcm coding standard," *ATT Tech. J.*, pp. 12–21, 1986. 1

[3] G. Frantz and R. Wiggins, "The development of "solid state speech" technology at texas instruments," in *Proc. IEEE Int. Conf. Acoust., Speech Signal Processing*, 1981. DOI: 10.1109/MSP.1981.28418. 1

[4] GSM, *GSM Digital Cellular Communication Standards: Enhanced Full-Rate Transcoding*, ETSI/GSM Std. GSM 06.60, 1996. 1

[5] M. Schroeder and B. Atal, "Code-excited linear prediction(CELP): High-quality speech at very low bit rates," in *Proc. IEEE Int. Conf. Acoust., Speech Signal Processing*, vol. 10, April 1985, pp. 937– 940. DOI: 10.1109/ICASSP.1985.1168147. 1

[6] K. Hellwig, P. Vary, D. Massaloux, J. Petit, C. Galand, and M. Rosso, "Speech codec for the european mobile radio system," in *Proc. of Global Telecom. Conf.*, vol. 2, November 1989, pp. 1065–1069. DOI: 10.1109/GLOCOM.1989.64121. 1

[7] G. D. Hair and T. W. Rekieta, "Automatic speaker verification using phoneme spectra," *J. Acoust. Soc. Amer.*, vol. 51, no. 1A, pp. 131–131, 1972. DOI: 10.1121/1.1981413. 2

[8] V. Berisha and A. Spanias, "Wideband speech recovery using psychoacoustic criteria," *EURASIP Journal on Audio, Speech, and Music Processing*, vol. 2007, 2007. DOI: 10.1155/2007/16816. 2, 5

[9] A. McCree, T. Unno, A. Anandakumar, A. Bernard, and E. Paksoy, "An embedded adaptive multi-rate wideband speech coder," in *Proc. IEEE Int. Conf. Acoust., Speech Signal Processing*, vol. 2, May 2001, pp. 761–764. DOI: 10.1109/ICASSP.2001.941026. 2, 24

[10] T. Unno and A. McCree, "A robust narrowband to wideband extension system featuring enhanced codebook mapping," in *Proc. IEEE Int. Conf. Acoust., Speech Signal Processing*, Philadelphia, PA, March 2005. DOI: 10.1109/ICASSP.2005.1415236. 2, 3, 4, 6, 13, 26, 32, 49, 52

[11] P. Jax and P. Vary, "Enhancement of band-limited speech signals," in *Proc. of Aachen Symposium on Signal Theory*, September 2001, pp. 331–336. DOI: 10.1109/ICASSP.1998.674450. 24, 30, 52

[12] P. Jax and P. Vary, "Artificial bandwidth extension of speech signals using MMSE estimation based on a hidden Markov model," in *Proc. IEEE Int. Conf. Acoust., Speech Signal Processing*, vol. 1, April 2003, pp. 680–683. DOI: 10.1109/ICASSP.2003.1198872. 3, 4, 24, 49, 52

[13] M. Nilsson and W. Kleijn, "Avoiding over-estimation in bandwidth extension of telephony speech," in *Proc. IEEE Int. Conf. Acoust., Speech Signal Processing*, vol. 2, May 2001, pp. 869–872. DOI: 10.1109/ICASSP.2001.941053. 3, 4, 13, 49, 52

[14] G. Chen and V. Parsa, "HMM-based frequency bandwidth extension for speech enhancement using line spectral frequencies," in *Proc. IEEE Int. Conf. Acoust., Speech Signal Processing*, vol. 1, May 2004, pp. 709–712. DOI: 10.1109/ICASSP.2004.1326084. 2

[15] Siyue Chen and Henry Leung, "Speech bandwidth extension by data hiding and phonetic classification," in *Proc. IEEE Int. Conf. Acoust., Speech Signal Processing*, vol. 4, April 2007, pp. 593–596. DOI: 10.1109/ICASSP.2007.366982. 2

[16] Chen, S. and Leung, H., "Artificial bandwidth extension of telephony speech by data hiding," in *Proc. IEEE Int. Symp. on Circuits and Systems*, May 2005, pp. 3151–3154. DOI: 10.1109/ISCAS.2005.1465296. 2

[17] V. Berisha and A. Spanias, "A scalable bandwidth extension algorithm," in *Proc. IEEE Int. Conf. Acoust., Speech Signal Processing*, vol. 4, April 2007, pp. 601–604. DOI: 10.1109/ICASSP.2007.366984. 2

[18] G. Geiser and P. Vary, "Backwards compatible wideband telephony in mobile networks: CELP watermarking and bandwidth extension," in *Proc. IEEE Int. Conf. Acoust., Speech Signal Processing*, vol. 4, April 2007, pp. 533–536. DOI: 10.1109/ICASSP.2007.366967. 6

[19] *An 8-32 kbit/s scalable wideband coder bitstream interoperable with G.729*, ITU-T Recommendation G.729.1, 2006. 6, 8, 24, 26, 32, 50, 53

[20] A. McCree, "A 14 kb/s wideband speech coder with a parametric highband model," in *Proc. IEEE Int. Conf. Acoust., Speech Signal Processing*, vol. 2, 2000. DOI: 10.1109/ICASSP.2000.859169. 2, 24

[21] P. Jax and P. Vary, "An upper bound on the quality of artificial bandwidth extension of narrowband speech signals," in *Proc. IEEE Int. Conf. Acoust., Speech Signal Processing*, vol. 1, May 2002, pp. 237–240. DOI: 10.1109/ICASSP.2002.5743698. 2, 4, 24, 49, 53

[22] M. Nilsson, and S.V. Anderson, and W.B. Kleijn, "On the mutual information between frequency bands in speech," in *Proc. IEEE Int. Conf. Acoust., Speech Signal Processing*, vol. 3, May 2000, pp. 1327–1330. DOI: 10.1109/ICASSP.2000.861823.

[23] M. Nilsson, and M. Gustafsson, and S.V. Anderson, and W.B. Kleijn, "Gaussian mixture model based mutual information estimation between frequency bands in speech," in *Proc. IEEE Int. Conf. Acoust., Speech Signal Processing*, vol. 1, May 2002, pp. 525–528. DOI: 10.1109/ICASSP.2002.5743770. 2, 4, 49, 53

[24] C. F. Chan and W. K. Hui, "Wideband re-synthesis of narrowband CELP coded speech using multiband excitation model," in *Proc. Int. Conf. on Spoken Language Processing*, vol. 1, Philadelphia, PA, October 1996, pp. 322–325. DOI: 10.1109/ICSLP.1996.607118. 3, 49

[25] V. Berisha and A. Spanias , "Enhancing the quality of coded audio using perceptual criteria," in *Proc. IEEE Wkshp. on Multimedia Signal Processing*, October 2005. DOI: 10.1109/MMSP.2005.248613. 3, 35, 36, 49

[26] V. Berisha and A. Spanias, "Enhancing vocoder performance for music signals," in *Proc. IEEE Int. Symp. on Circuits and Systems*, vol. 4, May 2005, pp. 4050–4053. DOI: 10.1109/ISCAS.2005.1465520.

[27] V. Berisha and A. Spanias, "Bandwidth extension of audio based on partial loudness criteria," in *Proc. IEEE Wkshp. on Multimedia Signal Processing*, October 2006. DOI: 10.1109/MMSP.2006.285286. 30, 49

[28] V. Berisha and A. Spanias, SPLIT-BAND Speech Compression Base on Loudness Estimation, ASU, Tempe, Arizona, US 8,392198 B1, 2012. 3

[29] H. Yasukawa, "Enhancement of telephone speech quality by simple spectrum extrapolation method," in *Proc. of EUROSPEECH*, September 1995, pp. 1545–1548. 3, 11, 13, 51, 52

[30] H. Yasukawa, "Signal restoration of broad band speech using nonlinear processing," in *Proc. of EUSIPCO*, September 1996, pp. 987–990. 3, 12, 23, 51, 52

[31] H. Yasukawa, "Wideband Speech Recovrery From Bandlimited Speech in Telephone Communications," in *Proc. IEEE Int. Symp. on Circuits and Systems*, no. 4, 1999, pp. 202–205. DOI: 10.1109/ISCAS.1998.698794. 3, 13, 51, 52

[32] P. Jax and P. Vary, *Audio Bandwidth Extension*. West Sussex, England: Wiley, 2005, ch. Bandwidth Extension for Speech, pp. 171–235. 4, 5, 13, 14, 15, 16, 17, 19, 28, 52

[33] H. Carl and U. Heute, "Bandwidth enhancement of narrow-band speech signals," in *Proc. of EUSIPCO*, vol. 2, September 1994, pp. 1178–1181. 4, 14, 16, 19, 30, 52

[34] Y. Yoshida and M. Abe, "An algorithm to reconstruct wideband speech from narrowband speech based on codebook mapping," in *Proc. Int. Conf. on Spoken Language Processing*, 1994, pp. 1591–1594. 4, 16, 52

64 BIBLIOGRAPHY

[35] Y.M. Cheng and D. O'Shaughnessy and P. Mermelstein, "Statistical recovery of wideband speech from narrowband speech," *IEEE Transactions on Speech and Audio Processing*, vol. 2, no. 4, pp. 544–548, 1994. DOI: 10.1109/89.326637. 4, 23, 24, 52

[36] S. Yao and C. F. Chan, "Block-based bandwidth extension of narrowband speech signals by using CDHMM," in *Proc. IEEE Int. Conf. Acoust., Speech Signal Processing*, Philadelphia, PA, March 2005. DOI: 10.1109/ICASSP.2005.1415233. 4, 24, 52

[37] Y. Nakatoh, M. Tsushima, and T. Norimatsu, "Generation of broadband speech from narrowband speech using piecewise linear mapping," in *Proc. of EUROSPEECH*, vol. 3, 1997, pp. 1643–1646. 4, 21, 52

[38] C. Avendano, and H.Hermansky, and E.A. Wan, "Beyond nyquist:towards the recovery of broad-bandwidth speech from narrow-bandwidth speech," in *Proc. of EUROSPEECH*, vol. 1, September 1995, pp. 165–168. 4, 20, 52

[39] J. Epps, "Wideband extension of narrowband speech for enhancement and coding," Ph.D. dissertation, The University of New South Wales, 2000. 6, 19, 21, 23, 53

[40] M. Dietz, and L. Liljeryd, and K. Kjorling, and O. Kunz, "Spectral band replication, a novel approach on audio coding," in *IEEE Aerospace and Electronic Systems*, 2002. 6, 8, 53

[41] J. S. Garofolo, and L. F. Lamel, and W. M. Fisher, and J. G. Fiscus, and D. S. Pallett, and N. L. Dahlgren, "The DARPA TIMIT acoustic-phonetic continuous speech corpus CD ROM," NTIS order number PB91-100354, Tech. Rep., February 1993. 6

[42] I. Maddieson, *Patterns of sounds.* Cambridge University Press, 1984. DOI: 10.1017/CBO9780511753459. 7

[43] P. Kroon and W.B. Kleijn, "Linear prediction-based analysis synthesis coding," in *Speech Coding and Synthesis*, 1995. DOI: 10.1007/978-1-4615-2281-2_3. 8, 35, 54

[44] H. Hermansky, "Perceptual linear predictive (PLP) analysis of speech," *J. Acoust. Soc. Amer.*, vol. 87, no. 4, pp. 1738–1752, April 1990. DOI: 10.1121/1.399423. 8, 54

[45] H. Strube, "Linear prediction on a warped frequency scale," *J. Acoust. Soc. Amer.*, vol. 68, pp. 1071–1076, 1980. DOI: 10.1121/1.384992. 8, 35, 54

[46] *Information Technology-Coding of Moving Pictures and Associated Audio for Digital Storage Media at up to about 1.5 Mbit/sec, IS11172-3: Audio*, ISO/IEC JTC1/SC29/WG11, 1992. 8, 35, 40, 42, 54, 55

[47] B. C. Moore, *An Introduction to the Psychology of Hearing*, 5th ed. New York: Academic Press, 2003. 8, 35, 43, 44, 54

[48] B. Moore, B. R. Glasberg, and T. Baer, "A model for the prediction of thresholds, loudness, and partial loudness," *J. Audio Eng. Soc.*, vol. 45, no. 4, 1997. 8, 9, 35, 38, 43, 44, 47, 50, 54, 55

[49] "The digital theater systems (dts)," web-page: www.dtsonline.com. 8, 35, 42, 55

[50] G. Davidson, *The Digital Signal Processing Handbook.* New York: CRC Press, 1998, ch. Digital Audio Coding: Dolby AC-3, pp. 41.1–41.21. 8, 35, 42, 55

[51] T. Painter and A. Spanias, "Perceptual segmentation and component selection for sinusoidal representations of audio," *IEEE Transactions on Speech and Audio Processing*, vol. 13, pp. 139–162, 2005. DOI: 10.1109/TSA.2004.841050. 9, 42, 55

[52] V. Atti and A. Spanias, "Speech analysis by estimating perceptually relevant pole locations," in *Proc. IEEE Int. Conf. Acoust., Speech Signal Processing*, vol. 1, March 2005, pp. 217–220. DOI: 10.1109/ICASSP.2005.1415089. 9, 35, 55

[53] H. Purnhagen, N. Meine, and B. Edler, "Sinusoidal coding using loudness-based component selection," in *Proc. IEEE Int. Conf. Acoust., Speech Signal Processing*, vol. 2, May 2002, pp. 1817–1820. DOI: 10.1109/ICASSP.2002.5744977. 9, 55

[54] L. Laaksonen and J. Kontio and P. Alku, "Artificial bandwidth expansion method to improve intelligibility and quality of AMR-coded narrowband speech," in *Proc. IEEE Int. Conf. Acoust., Speech Signal Processing*, vol. 1, March 2005, pp. 809–812. DOI: 10.1109/ICASSP.2005.1415237. 12, 23

[55] J. Paulus, "Variable rate wideband speech coding using perceptually motivated thresholds," in *IEEE Speech Coding Workshop*, 1995, pp. 35–36. DOI: 10.1109/SCFT.1995.658114. 14

[56] C. Erdmann, P. Vary, K. Fischer, W. Xu, M. Marke, T. Fingscheidt, I. Varga, M. Kaindl, C. Quinquis, B. Kovesi, and D. Massaloux, "A candidate proposal for a 3GPP adaptive multi-rate wideband speech codec," in *Proc. IEEE Int. Conf. Acoust., Speech Signal Processing*, vol. 2, May 2001, pp. 757–760. DOI: 10.1109/ICASSP.2001.941025. 27

[57] *AMR-WB; Transcoding Functions*, 3GPP TS 26.190, 2003. 14, 21, 23, 26, 27, 51

[58] Y. Linde, A. Buzo, and R. Gray, "An algorithm for vector quantizer design," *IEEE Trans. on Comm.*, vol. 28, pp. 84–95, 1980. DOI: 10.1109/TCOM.1980.1094577. 17, 18

[59] ITU, "Coding of speech at 8kb/s using conjugate-structure algebraic-code-excited linear-prediction (cs-acelp)," no. Draft Recommendation G.729, 1995. 18

[60] J. Conway and N. Sloane, "Voronoi regions of lattices, second moments of polytopes, and quantization," *IEEE Trans. on Inf. Theory*, vol. 28, 1982. DOI: 10.1109/TIT.1982.1056483. 19

[61] S. Chennoukh, A. Gerrits, G. Miet, and R. Sluijter, "Speech enhancement via frequency bandwidth extension using line spectral frequencies," in *Proc. IEEE Int. Conf. Acoust., Speech Signal Processing*, vol. 1, May 2001, pp. 665–668. DOI: 10.1109/ICASSP.2001.940919. 20, 21

[62] J. Cabral and L. Oliveira, "Pitch-synchronous time-scaling for high-frequency excitation regeneration," in *Proc. of Interspeech*, 2005. 26, 31

[63] D. Griffin and J. Lim, "Multiband excitation vocoder," *IEEE Transactions on Speech and Audio Processing*, vol. 37, no. 4, pp. 246–275, 1988. DOI: 10.1109/29.1651. 27

[64] R.J. McAulay and T.F. Quatieri, *Speech Coding and Synthesis*. Elsevier, 1995, ch. Sinusoidal Coding, pp. 121–173. 27

[65] R. McAulay and T. Quatieri, *Speech Coding and Synthesis*. Elsevier, 1995, ch. Time-domain and frequency-domain techniques for prosodic modification of speech, pp. 519–555. 28

[66] W. Verhelst and M. Roelands, "An overlap-add technique based on waveform similarity (WSOLA) for high quality time-scale modification of speech," in *Proc. IEEE Int. Conf. Acoust., Speech Signal Processing*, vol. 2, April 1993, pp. 554–557. DOI: 10.1109/ICASSP.1993.319366. 28

[67] U. Kornagel, "Spectral widening of the excitation signal for telephone-band speech enhancement," in *Proc. Int. Wkshp. on Acoust. Echo and Noise Control*, 2001, pp. 215–218. 30

[68] Y. Qian and P. Kabal, "Combining equalization and estimation for bandwidth extension of narrowband speech," in *Proc. IEEE Int. Conf. Acoust., Speech Signal Processing*, vol. 1, May 2004, pp. 713–716. DOI: 10.1109/ICASSP.2004.1326085. 32, 34

[69] V. Atti and A. Spanias, "Rate determination based on perceptual loudness," in *Proc. IEEE Int. Symp. on Circuits and Systems*, vol. 2, May 2005, pp. 848–851. DOI: 10.1109/IS-CAS.2005.1464721. 35

[70] A. Spanias, T. Painter, and V. Atti, *Audio Signal Processing and Coding*. New York: Wiley-Interscience, 2006. 35, 39, 41

[71] B.C.J. Moore and B.R. Glasberg, "Audibility of time-varying signals in time-varying backgrounds: Model and data," *J. Acoust. Soc. Amer.*, vol. 115, pp. 2603–2603, May 2001. DOI: 10.1121/1.1738839. 35

[72] B.C.J. Moore and B.R. Glasberg, "A model of loudness applicable to time-varying sounds," *J. Audio Eng. Soc.*, vol. 50, pp. 331–342, May 2002. 35

[73] E. Zwicker and H. Fastl, *Psychoacoustics*. Springer, New York, 1990. 35, 38, 42, 43

[74] ITU, "Perceptual evaluation of speech quality (PESQ): An objective method for end-to-end speech quality assessment of narrow-band telephone networks and speech codecs," no. Recommendation 862, 1995. 36

[75] B. Paillard, P. Mabilleau, S. Morissette, and J. Soumagne, "PERCEVAL: perceptual evaluation of the quality of audio signals," *J. Acoust. Soc. Amer.*, vol. 40, no. 1-2, pp. 21–31, 1992. 36

[76] A. Tewfik, M. Swanson, and B. Zhu, "Data embedding in audio: Where do we stand?" in *Proc. IEEE Int. Conf. Acoust., Speech Signal Processing*, March 1999, p. 2075. DOI: 10.1109/ICASSP.1999.758340. 36

[77] E. Vickers, "Automatic long-term loudness and dynamics matching," in *Proc. of Audio Eng. Soc. Conv.*, September 2001. 36

[78] H. Fletcher, "Perceptual segmentation and component selection for sinusoidal representations of audio," *Revs of Modern Physics*, vol. 12, pp. 47–65, 1940. DOI: 10.1103/RevModPhys.12.47. 36, 37, 38

[79] T. Painter and A. Spanias, "Perceptual coding of digital audio," in *Proc. of the IEEE*, vol. 88, April 2000, pp. 451–515. DOI: 10.1109/5.842996. 38, 39, 40, 41

[80] B. Moore and B. R. Glasberg, "Formulae describing frequency selectivity as a function of frequency and level, and their use in calculating excitation patterns," *Hearing Research*, pp. 209–225, 1987. DOI: 10.1016/0378-5955(87)90050-5. 42, 43

[81] B. Edler and G. Schuller, "Audio coding using a psychoacoustic pre- and post-filter," in *Proc. IEEE Int. Conf. Acoust., Speech Signal Processing*, 2000, pp. 881–884. DOI: 10.1109/ICASSP.2000.859101. 49

[82] *AMR-NB; Transcoding Functions*, 3GPP TS 26.090, 2001. 51

[83] F. Nagel and S. Disch, "A harmonic bandwidth extension method for audio codecs," in *Acoustics, Speech and Signal Processing, 2009. ICASSP 2009. IEEE International Conference on*, 2009, pp. 145–148. DOI: 10.1109/ICASSP.2009.4959541. 57

[84] F. Nagel, S. Disch, and S. Wilde, "A continuous modulated single sideband bandwidth extension," in *Acoustics Speech and Signal Processing (ICASSP), 2010 IEEE International Conference on*, 2010, pp. 357–360. DOI: 10.1109/ICASSP.2010.5495843. 57

[85] G. Alipoor and M. Savoji, "Wide-band speech coding based on bandwidth extension and sparse linear prediction," in *Telecommunications and Signal Processing (TSP), 2012 35th International Conference on*, 2012, pp. 454–459. DOI: 10.1109/TSP.2012.6256335. 57

[86] N. Park, Y. Lee, and H. Kim, "Artificial bandwidth extension of narrowband speech signals for the improvement of perceptual speech communication quality," in *Communication and Networking*, ser. Communications in Computer and Information Science, T.-h. Kim, H. Adeli, W.-c. Fang, T. Vasilakos, A. Stoica, C. Patrikakis, G. Zhao, J. Villalba, and Y. Xiao, Eds. Springer Berlin Heidelberg, 2011, vol. 266, pp. 143–153. [Online]. Available: http://dx.doi.org/10.1007/978-3-642-27201-1_17 DOI: 10.1007/978-3-642-27201-1_17. 57

[87] C. Yağli, M.A. Tuğtekin Turan, and Engin Erzin, "Artificial bandwidth extension of spectral envelope along a Viterbi path," *Speech Communication*, vol. 55, no. 1, pp. 111 – 118, 2013. [Online]. Available: http://www.sciencedirect.com/science/article/pii/S0167639312000957 DOI: 10.1016/j.specom.2012.07.003. 57

[88] L. Laaksonen and J. Virolainen, "Binaural artificial bandwidth extension (b-abe) for speech," in *Acoustics, Speech and Signal Processing, 2009. ICASSP 2009. IEEE International Conference on*, 2009, pp. 4009–4012. DOI: 10.1109/ICASSP.2009.4960507. 57

[89] J. Li, D. Yu, J.-T. Huang, and Y. Gong, "Improving wideband speech recognition using mixed-bandwidth training data in cd-dnn-hmm," in *Spoken Language Technology Workshop (SLT), 2012 IEEE*, 2012, pp. 131–136. DOI: 10.1109/SLT.2012.6424210. 57

[90] H. Pulakka and P. Alku, "Bandwidth extension of telephone speech using a neural network and a filter bank implementation for highband mel spectrum," *Audio, Speech, and Language Processing, IEEE Transactions*, vol. 19, no. 7, pp. 2170–2183, 2011. DOI: 10.1109/TASL.2011.2118206. 57

[91] H. Pulakka, U. Remes, S. Yrttiaho, K. Palomaki, M. Kurimo, and P. Alku, "Bandwidth extension of telephone speech to low frequencies using sinusoidal synthesis and a gaussian mixture model," *Audio, Speech, and Language Processing, IEEE Transactions*, vol. 20, no. 8, pp. 2219–2231, 2012. DOI: 10.1109/TASL.2012.2199110. 57

[92] I. Katsir, D. Malah, and I. Cohen, "Evaluation of a speech bandwidth extension algorithm based on vocal tract shape estimation," in *Acoustic Signal Enhancement; Proceedings of IWAENC 2012; International Workshop on*, 2012, pp. 1–4. 57

[93] Y. Kwon, Y. Li, and S. Kang, "Bandwidth extension of g.729 speech coder using search-free codebook mapping," in *Telecommunications and Signal Processing (TSP), 2012 35th International Conference on*, 2012, pp. 437–440. DOI: 10.1109/TSP.2012.6256331. 57

[94] P. Heewan, K. Sangwon, and A. Spanias, "Search-free codebook mapping for artificial bandwidth extension," *IEICE Transactions on Communications*, vol. 95, no. 4, pp. 1479–1482, 2012. DOI: 10.1587/transcom.E95.B.1479. 57

[95] K.N. Ramamurthy and A.S. Spanias, MATALB Software for the Code Excited Linear Prediction Algorithm: The Federal Standard-1016, Morgan and Claypool Publishers, Vol. 2, No. 1, ISBN 1608453847, Jan. 2010 1

[96] J.J. Thiagarajan and Andreas Spanias, Analysis of MPEG-1 Layer III (MP3) Algorithm Using MATLAB, Morgan and Claypool Publishers, Vol. 3, No. 3, ISBN 10:1608458016, ISBN 13:1608458011, November 2011. 39

Authors' Biographies

VISAR BERISHA

Visar Berisha is an Assistant Professor with a joint appointment in the Department of Speech and Hearing Science and the School of Electrical, Computer, and Energy Engineering at Arizona State University. His research interests fall mainly in the field of speech and audio perception, signal processing, and machine learning. He obtained his Ph.D. in Electrical Engineering at Arizona State University. Following his degree, he worked at MIT Lincoln Laboratory and Raytheon Co. as research engineer.

STEVEN SANDOVAL

Steven Sandoval received a B.S. Electrical Engineering in 2007 and his M.S. Electrical Engineering in 2010 from the Klipsch School of Electrical and Computer Engineering, New Mexico State University, Las Cruces, NM. He previously worked for five years as a system analyst for a defense contractor. He is presently working on his Ph.D. degree in electrical engineering in the Ira A. Fulton Schools of Engineering, SenSIP Center, Arizona State University, Tempe, AZ. His research interests include signal processing, specifically audio and speech processing, time-frequency analysis, machine learning, and robotics.

JULIE LISS

Julie Liss is a Professor in the Department of Speech and Hearing Science. She is Director of the Motor Speech Disorders Lab where she conducts research on the perception of degraded speech. Her interest is in modeling the cognitive-perceptual strategies involved in deciphering degraded speech to further elucidate the construct of speech intelligibility. She and Dr. Berisha work collaboratively to apply signal processing methodology to issues intelligibility in clinical populations.

Printed in the United States
by Baker & Taylor Publisher Services